87

Structure and Bonding

Springer-Verlag Berlin Heidelberg GmbH

Structural and Electronic Paradigms in Cluster Chemistry

Volume Editor: D. M. P. Mingos

With contributions by
J. D. Corbett, M.-F. Fan, T. P. Fehlner, J.-Halet,
C. E. Housecroft, R. L. Johnston, Z. Lin, J.-Y. Saillard

Springer

In references Structure and Bonding is abbreviated
Struct. Bond. and is cited as a journal.

Springer WWW home page: http://www.springer.de

ISSN 0081 - 5993
ISBN 978-3-662-14816-7

Die Deutsche Bibliothek - CIP-Einheitsaufnahme
Structural and electronic paradigms in cluster chemistry /
vol. ed.: D. M. P. Mingos. With contributions by J. D. Corbett ... -
(Structure and bonding ; 87)
ISBN 978-3-662-14816-7 ISBN 978-3-540-68689-7 (eBook)
DOI 10.1007/978-3-540-68689-7

CIP Data applied for

Originally published by Springer-Verlag Berlin Heidelberg New York in 1997
Softcover reprint of the hardcover 1st edition 1997

Typesetting: Fotosatz-Service Köhler OHG, Würzburg
Cover: Medio V. Leins, Berlin
SPIN: 10552960 66/3020 - 5 4 3 2 1 0 - Printed on acid-free paper

Volume Editor

Prof. D. M. P. Mingos
Chemistry Department
Imperial College of Science Technology and Medicine
South Kensington
London SW7 2AY, Great Britain
E-mail: d.mingos@ic.ac.uk

Editorial Board

Preface

The last five decades have witnessed a spectacular growth in the chemistry of polyhedral molecules of the elements.[1)] In the main group area research which started in the Cold War era with the synthesis of new boranes and carboranes revealed a whole class of symmetrical molecules which had deltahedral polyhedral structures, i.e. $B_nH_n^{2-}$ and $C_2B_{n-2}H_n$. These compounds have been joined subsequently by many examples of isoelectronic molecules containing other main group metals and nonmetals, e.g. Sn_9^{4-} Bi_9^{5+}, which also have polyhedral cage structures but do not have any extra-cage bonds to hydrogen atoms. Some indications of the possibilities for cage molecules of this type were evident in the earlier German literature which suggested that salts such as NaPb actually were better formulated as Na_4Pb_4, with tetrahedral polyhedral because of their isoelectronic relationships with P_4, As_4, etc. This and subsequent research firmly established the possibility of making "bare" clusters of the main group elements either as allotropes of the element or as cationic and anionic derivatives. Recent developments in this field are summarised in Professor Corbett's chapter. The isolation of such cationic and anionic polyhedral ions on a reasonable scale required chemists to do their synthesis in solvent media either of low nucleophilicity or electrophilicity. More recently it has been recognised that such species could also be generated and studied in a mass spectrometry chamber on a molecular scale using molecular beam techniques. It was such experiments which lead to the spectacular development of C_{60} chemistry which was recognised in 1996 by the award of the Nobel Prize jointly to Smalley, Curl and Kroto.

The in-depth study of transition metal cluster compounds evolved out of a curiosity towards the valence problems posed by their novel-structural features and a belief that small molecular clusters could prove to be effective homogenous catalysts for the conversion of CO and H_2 into hydrocarbons and alcohols. Initially cluster compounds with only 3–6 metal atoms were isolated, but the improvements in spectroscopic, structural and separation techniques rapidly led to the characterisation of many high nuclearity cluster compounds and there now exist examples of structurally well defined compounds with more than 100 metal atoms. The study of cluster compounds was also stimulated by the superconductivity properties of the Chevrel phase molybdenum-sulfido octahedral clusters and the recognition that molybdenum–iron–sulfido clusters may be involved in the active site of nitrogenase.

These experimental findings have been matched by the development of a conceptual framework which have brought some intellectual order to the multitude of polyhedral structures revealed in main groups and transition metal chemistry. A detailed account of the historical development of these ideas and their widespread applications have been given elsewhere and will not be repeated here.[2] The purpose of this Preface is to bring attention to the key papers which led to the development of the Polyhedral Skeletal Electron Pair Theory.

In 1971 Bob Williams[3] recognised for the first time that the structures of the boranes B_nH_{n+4} and B_nH_{n+6} were not all based on fragments of an icosahedron, but more correctly viewed as almost complete Skeletons of the parent deltahedral boranes $B_nH_n^{2-}$. Specifically he established triads of molecules which were structurally related and had either the complete deltahedral cage, (the *closo*-$B_nH_n^{2-}$ anions), one vertex missing from the complete cage (*nido*-B_nH_{n+4}) or two vertices missing from the complete cage (*arachno*-B_nH_{n+6}). In the same year Ken Wade recognised the validity of this structural relationship and proposed a simple electronic basis for it. He developed the important principle that the *closo*-, *nido*- and *arachno*-triads were structurally related because they had the same number of electron pairs involved in skeletal bonding. Specifically the (n+1) electron pairs characteristic of the closoboranes $B_nH_n^{2-}$ were retained in the *nido*- and *arachno*-derivatives $[B_{n-1}H_{(n-1+4)}]$ and $[B_{n-2}H_{(n-2+6)}]$.[4] Wade also recognised that the bonding pattern observed in *closo*-, *nido*- and *arachno*-borane polyhedral molecules were also reproduced in metal carbonyl clusters. Professor Fehlner's Chapter on metalloboranes illustrates how important this relationship proved to be. These ideas were later extended to electron precise and electron rich and capped clusters in 1972 by Wade[5] and myself.[6] These papers also recognised the importance of interstitial atoms in cluster chemistry and the Chapter by Dr. Housecroft summarises the current methodologies used to accommodate interstitial atoms with the electron counting generalisations.

The theoretical basis of the electron counting rules was underpinned during the subsequent decade. The cluster compounds of the early transition metals could be incorporated within the framework of the Polyhedral Skeletal Electron Pair Theory, but it was recognised that the spectrum of molecular orbitals generated in such clusters differed from those in π-acceptor clusters. The Chapter by Zhenyang Lin and Man-Fai Pan demonstrates how the great majority of π-donor clusters may be incorporated within the general framework of the approach.

The conceptual models which were first introduced 25 years ago proved to be important for several reasons. Firstly, they provided a structural paradigm which could be used to rationalise the structures of polyhedral molecules in an analagous way to that which had been achieved two decades earlier for mononuclear compounds in the *Valence Shell Electron Pair Repulsion Theory*.[8] Secondly it unified for the first time the areas of main group and transition metal cluster chemistry. The relationship between main group and transition metal carbonyl fragments were formalised in the *isolobal* analogy in 1976.[9, 10] Thirdly the research established that the closed shell requirements of even these complex molecules were governed by relatively simple electronic requirements. In the 1980s Stone was able to demonstrate that the closed shell requirements of *closo*-, *nido*- and *arachno*-polyhedral molecules could be rationalised within the framework of a spherical free electron model. The Tensor Surface Harmonic model and its group theoretical implications have provided an elegant way of accounting for many features of the electron counting rules developed earlier.[11] Dr. Johnston's Chapter on the mathematical aspects of cluster chemistry provides an excellent introduction to these fundamental aspects.

Although Polyhedral Skeletal Electron Pair theory has survived the test of time, many new applications have developed and further theoretical work has hightlighted the reasons for apparent exceptions to the generalisations. The Chapter by Drs Halet and Saillard provides a good illustration of how some of these exceptions have been rationalised using modern theoretical techniques.

I hope that collectively the Chapters provide a timely celebration of the seminal contributions of Williams and Wade and a contemporary account of some of the more important recent developments. Their perception and imagination opened up a new area of chemistry and inspired us all.

London
March 1997

D. M. P. Mingos

1. Mingos DMP, Wales DJ (1990) Introduction to cluster chemistry, Prentice-Hall, New York
2. Johnston RL, Mingos DMP (1987): Struct Bond, 29, 68
3. Williams RE (1971) Inorg Chem, 10, 210
4. Wade KJ (1972) Chem Soc Chem Commun 792
5. Wade K (1972) Inorg Nucl Chem Lett 8, 559, 563, 823
6. Mingos DMP (1972) Nature Phys Sci 99, 236
7. Mason R, Thomas KM, Mingos DMP (1973) J Amer Chem Soc 96, 3802
8. Gillespie RJ, Nyholm RS (1957) Quart Rev Chem Soc 11, 339
9. Elian M, Chen MML, Mingos DMP, Hoffmann R (1970) Inorg Chem 15, 1148
10. Hoffmann R (1985) Nobel Lecture, Science, 211, 995
11. Stone AJ (1980) Mol Phys 1, 1339; (1981) Inorg Chem 20, 563

Contents

Contents of Volume 86

Atoms and Molecules in Intense Fields

Volume Editors: L. S. Cederbaum, K. C. Kulander, N. H. March

Mathematical Cluster Chemistry

Roy L. Johnston

School of Chemistry, University of Birmingham, Edgbaston, Birmingham B15 2TT, United Kingdom. *E-mail: R.L.Johnston.che@bham.ac.uk*

The application of mathematical methods, such as polyhedral topology, graph theory and group theory, to the study of cluster molecules is reviewed. The development of localised orbital and spherical harmonic methods are outlined and there is an introduction to Tensor Surface Harmonic Theory and Group Theory on a Spherical Shell. The inter-relationships between these approaches and their links with simple molecular orbital models, such as Hückel Theory, are detailed and rationalised. The review concludes with case studies which show how some or all of these methodologies have been brought to bear on three problems - deviations from Wade's rules in "non-spherical" borane clusters, the occurrence of approximately non-bonding skeletal molecular orbitals in hydrocarbon clusters; and the prediction of the structures, electronic properties and stabilities of the fullerenes - truly a "mathematical chemist's delight".

Keywords: Clusters; graph theory; group theory; spherical harmonics; topology

1
Introduction to Mathematical Cluster Chemistry

Since the first cluster molecules were structurally characterised over 30 years ago, there has been tremendous interest in trying to rationalise their geometries and, in particular, in relating their geometric and electronic structures [1]. The polyhedral nature of these clusters, which often have high symmetries, has led theoreticians to apply a number of mathematical methodologies to their study. These methods include: polyhedral geometry and topology; graph theory; group theory; and spherical harmonic treatments. There are many links between these fields. Graph theory, for example, is a branch of topology and spherical harmonic treatments involve group theory in the spherical group R_3. One of the major contributors to the field of mathematical (cluster and coordination) chemistry over the past 20 years has been R. B. King. For an overview of his work in the area of mathematical inorganic chemistry the reader is referred to a recent review [2].

The success of topological and group theoretical approaches, which concentrate on the cluster polyhedron without being concerned with the actual nature of the cluster atoms, has led to the realisation that most cluster geometry-electron count correlations are symmetry and topology dependent. This also explains the success of simple Molecular Orbital (MO) treatments, such as Hückel theory, and symmetry-based semi-empirical MO methods, e.g. the extended Hückel method, in rationalising and predicting cluster geometries and electron counts [3].

This review will not cover the many applications of MO theory to main group and transition metal cluster compounds and the important empirical and phenomenological observations and generalisations that have been made on the basis of these calculations. For an overview of the innovative and seminal work of Wade [4, 5], Mingos [6, 7] and others the reader is referred to other reviews in this volume, to a previous review by Mingos and Johnston [1], and to the book by Mingos and Wales [8].

2 Clusters as Polyhedra

Polyhedra, their shapes, symmetries, and mathematical properties have been a source of fascination to scientists throughout the ages – from Euclid (or even earlier) to the present day [9]. The properties and interrelationships between polyhedral structures, and some of their applications (or occurrences) in nature, engineering and science can be found in a fascinating book by Williams [10].

Over the past quarter of a century, the synthesis and structural characterisation of cluster molecules, with a variety of polyhedral structures, has led many theoretical chemists to explore the mathematics of polyhedral geometry and topology. Though much of this work has involved the rediscovery of geometrical rules and constructions which may be found in the mathematical literature, a number of methods and derivations have been presented for the first time [11, 12]. In my opinion the study of cluster geometry, and how it is related to the electronic structures of clusters, has put new life into the field of polyhedral geometry and topology. Some of the most significant contributions to the field will be mentioned later in this review.

2.1 Polyhedral Topology: Euler's Relations

I shall limit my discussion to 3-dimensional polyhedral clusters whose atoms lie approximately on the surface of a single sphere and which are convex (i.e. all polygons are convex).

The best known polyhedra are the five regular convex polyhedra (the "Platonic solids": the tetrahedron, octahedron, cube, icosahedron and dodecahedron), which have the following properties [9, 10]:

i) all polygons (faces) are regular and equivalent;
ii) all vertices are equivalent;
iii) all edges are equivalent;
iv) all dihedral angles (formed by two faces meeting at a common edge) are equal.

It is worth noting that these polyhedra have all been realised as framework (skeletal) structures for main group and/or transition metal clusters. A variety of other types of polyhedra can be generated by relaxing some or all of these constraints. The cuboctahedron, for example, is an example of a "quasi-regular" polyhedron, in that it possesses both square and triangular faces, though all other restrictions apply [9, 10].

There are a number of rules relating the topological features – i.e. the number and types of vertices, edges and faces – of polyhedra. Some of these are summarised below [9, 10].

The fundamental equation governing polyhedral topology is "Euler's relation" which, for 3-D convex polytopes, takes the form:

$$V + F = E + 2 \tag{1}$$

where V, F and E are the total numbers of vertices, faces and edges of the polyhedron respectively (note that in later sections of this review, V is replaced by n, the number of skeletal cluster atoms).

The octahedron, for example, is characterised by: V (6) + F (8) = E (12) + 2.

Equation (1) can be used to generate the following fundamental relationships [13]:

$$\Sigma_k f_k - {}^1/_2\, \Sigma_j (j\text{-}2)\, v_j = 2 \tag{2a}$$

$$\Sigma_j v_j - {}^1/_2\, \Sigma_k (k\text{-}2)\, f_k = 2 \tag{2b}$$

where there are v_j vertices of connectivity j, f_k faces with k sides, and the total number of vertices and faces are given by: $V = \Sigma_j v_j$; $F = \Sigma_k f_k$. Equations (2a) and (2b) can be combined to give another useful expression of the Euler relationship:

$$\Sigma_j\, j \cdot v_j = \Sigma_k\, k \cdot f_k = 2E \tag{3}$$

2.2 Types of Polyhedra

A number of polyhedral structure-types are commonly found in cluster chemistry. Some of these, and their properties, are summarised below.

(i) Trivalent (or 3-connected) polyhedra – all vertices have three neighbours (i.e. three edges). Substitution of j = 3 into Eq. (2) yields E = 3V/2 and substitution into Eq. (1) gives F = V/2 + 2. These polyhedra (some examples of which are shown in Fig. 1) are of great relevance to cluster chemistry and will be discussed in greater detail in Sect. 7.

(ii) Deltahedra – all faces are triangular. Substitution of k = 3 into Eq. (2) yields E = 3F/2 and substitution into Eq. (1) gives V = F/2 + 2. These polyhedra are of great significance to cluster chemistry, since they correspond to the structures adopted by the *closo*-boranes $[B_nH_n]^{2-}$.

(iii) Tetravalent (or 4-connected) polyhedra – all vertices have four neighbours and, by analogy with the trivalent polyhedra: E = 2V; F = V + 2.

(iv) Rhombohedra – all faces are 4-sided squares or rhomboids. By analogy with the deltahedra: E = 2F; V = F + 2.

Tetravalent polyhedra and rhombohedra are less common as cluster geometries, though it should be noted that the octahedron is tetravalent as well as being deltahedral, and the cube is a rhombohedron as well as being trivalent. Similarly, the tetrahedron is a trivalent deltahedron.

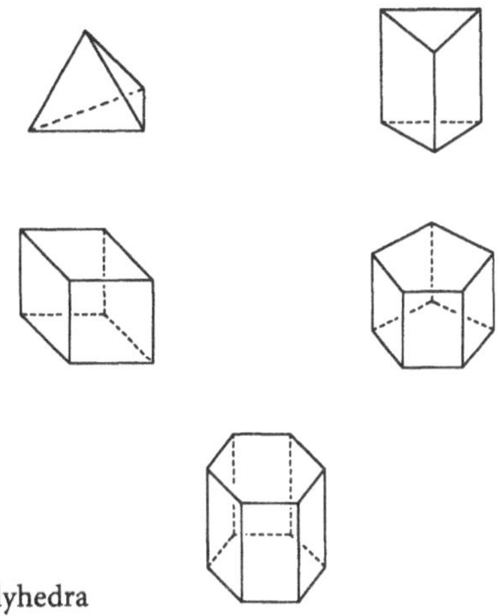

Fig. 1. Some trivalent polyhedra

2.3 Duality

It is clear from (i)–(iv) above that there are relationships between the topological indices (numbers of vertices, edges and faces) of these polyhedra. Thus, for a trivalent polyhedron, $F = V/2 + 2$, while for a deltahedron $V = F/2 + 2$. These relationships are due to the fact that trivalent polyhedra and deltahedra are "Geometric Duals".

The geometric dual (D) of a general polyhedron P is constructed by placing a vertex at the centroid of each of the faces of the polyhedron and connecting vertices which correspond to faces sharing a common edge, as shown in Fig. 2.

In this way, a k-sided face in polyhedron P becomes a k-connected vertex in the dual D. The number of vertices of D (V_D) is therefore equal to the number of faces of P (F_P). The dual relationship is reciprocal so applying the dual construction to D generates the original polyhedron P. Thus, the number of faces of D (F_D) is equal to the number of vertices of P (V_P). Substitution into Eq. (1) reveals that polyhedra P and D have the same number of edges (i.e. $V_P + F_P = F_D + V_D = E + 2$). These relationships are listed in Table 1.

From the above definitions, it is apparent that trivalent polyhedra and deltahedra constitute dual series of polyhedra as do the tetravalent polyhedra and rhombohedra. The octahedron (tetravalent and deltahedral) and the cube (trivalent and rhombohedral) serve as an illustration of both relationships: the

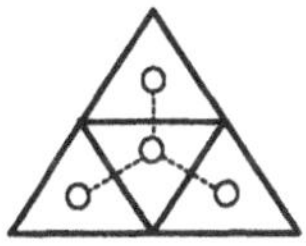

Fig. 2. Construction of the dual (*dashed lines*) of a polyhedron (*bold lines*) by joining the centroids (*white circles*) of edge-sharing faces

Table 1. Relationships between topological features of dual polyhedra

	P v_j j-conn. vertices f_k k-sided faces	D $f_j = v_j$ j-sided faces $v_k = f_k$ k-conn. vertices
Vertices (V)	V_P	$V_D = F_P$
Faces (F)	F_P	$F_D = V_P$
Edged (E)	E_P	$E_D = E_P$

octahedron has six tetravalent vertices and eight 3-sided faces; the cube (its dual) has eight trivalent vertices and six 4-sided faces.

Another important concept is that of the "geometric edge-figure" of a polyhedron (or "line graph" in the terminology of graph theory [14, 15]). The edge-figure (E) of polyhedron P is generated by placing vertices at the midpoints of the edges of P and connecting those vertices which correspond to coincident edges in P, provided that these edges bound the same face of the original polyhedron P. Since polyhedra always have more edges than vertices, the edge-figure relationship is not a reciprocal process (i.e. the edge-figure of E is not P).

It has been shown above that any pair of dual polyhedra P and D have the same number of edges. It can also be shown that these polyhedra have the same edge-figure. Another way of generating E is to construct P and D so that their edges are tangential to the surface of a common sphere. The vertices of E are given by the points of intersection of the edges of P and D and lie on the sphere as they must correspond to the tangent points. Further consideration of the edge-figure construction proves that E is always a tetravalent polyhedron since, in any polyhedron, a specified edge can only be coincident with two other edges (at each end) which bound faces in common with the first edge.

In this section, I have established some of the geometrical rules and topological relationships between polyhedra. Applications of these concepts to the study of cluster bonding will be discussed in later sections.

3 Clusters as Graphs

Graph theory has long been of interest to theoretical cluster chemists [2]. In fact, every time we draw a 2-D representation of a cluster, we are effectively describing the cluster as a graph [14, 15]. The relationship between a simple line drawing of a cluster and its electronic structure can be appreciated by making the connection between graph theory and simple MO theory.

Following King [2], the connectivity or "adjacency" matrix (A) of an n-vertex cluster graph *G* is an $n \times n$ matrix with elements A_{ij} (=1 if vertices i and j are connected by an edge; = 0 if i = j or i and j are not connected). The eigenvalues (x_k) of A are solutions of the equation

$$|A\text{-}xI| = 0 \tag{4}$$

where I is the unit matrix.

3.1
Graph Theory and Hückel Theory

There is a well known [2, 16, 17] topological equivalence between the adjacency matrix (A) of graph theory and the Hamiltonian matrix (H) of Hückel MO theory [18], since in Hückel theory: $H_{ii} = \alpha$; $H_{ij} = \beta$ (i and j bonded); and $H_{ij} = 0$ (i and j not bonded). In Hückel theory, α and β are known as the Coulomb and resonance integrals respectively and give a measure of the energies of and strength of interaction between the atomic orbitals. The eigenvalues (MO energies) of Hückel theory are related to those of graph theory as follows:

$$\varepsilon_k = \alpha + x_k\beta \tag{5}$$

in the limit of zero overlap between atomic orbitals. This topological equivalence underpins the application of graph theory to cluster bonding [17, 19].

Since the Hückel MOs are often a reasonable approximation to the ab initio wavefunctions, and the orbital orderings (if not the exact energies) are well reproduced, graph theory enables the rationalisation of the results of numerically accurate (though often conceptually obscure) MO calculations in a simple and transparent way. A recent development of Hückel theory has been made by Gimarc in his 3-dimensional Hückel method for calculating the π-symmetry surface MOs of main group clusters [20].

3.2
Graph Theory and Moments

In the late 1960s, Cyrot-Lackman showed that the moments of an eigenspectrum can be related to the topology of the system under study [21]. The "method of moments" has been used widely in solid state physics: in solids the discrete eigenspectrum is replaced by the continuous density of states (DOS) and has been successfully applied to the study of inorganic solids and clusters, notably by Burdett [22] and Rousseau and Lee [23].

The pth moment of an eigenspectrum (of MOs), μ_p is defined as

$$\mu_p = \Sigma_n(\varepsilon_n)^p \tag{6}$$

where n runs over all occupied and unoccupied orbitals. Since, by definition, the Hamiltonian matrix (H) is diagonal with respect to the set of eigenfunctions $\{\Psi_n\}$; $H_{nm} = \varepsilon_n\delta_{nm}$, Eq. (6) can be rewritten

$$\mu_p = \Sigma_n(H_{nn})^p = \Sigma_n(H^p)_{nn} = \mathrm{Tr}(H^p) \tag{7}$$

The trace of the Hamiltonian matrix (to the pth power) is invariant with respect to the choice of basis set, if these are related by a unitary transformation, so that the moments can be given in terms of the basis set of atomic orbitals $\{\phi_i\}$:

$$\mu_p = \mathrm{Tr}(H^p) = \Sigma_i(H^p)_{ii} = \Sigma_i(H_{ii})^p \tag{8}$$

where $H = U^{-1}HU$. Performing the matrix multiplication explicitly:

$$\mu_p = \Sigma_{i1,i2\ldots ip}\, H_{i1,i2}\, H_{i2,i3} \ldots H_{ip,i1} \tag{9}$$

the pth moment of the eigenspectrum is seen to be equal to the sum over all paths of length p that start and end on the same atom. Thus, by rendering the cluster as a graph and making the Hückel approximations, the relationship between the topology of the cluster and the various moments of the eigenspectrum can be obtained. This enables the width and shape of the spectrum of MOs of a cluster to be related to topological features, such as the number of edges and faces of a given size (e.g. deltahedral clusters will have large values of μ_3 as there are many paths of length 3 around the triangular faces). The nearest neighbour nature of the Hückel Hamiltonian dramatically cuts down on the number of paths contributing to μ_p, since all the steps in an allowed path must be between neighbouring (i.e. "bonded") atoms, or else the product $H_{i1,i2} H_{i2,i3} \ldots H_{ip,i1}$ vanishes.

The effect of the lower moments on the appearance of the MO spectrum are as follows:

$$\mu_0 = \mathrm{Tr}(H^0) = N \tag{10}$$

(the number of eigenstates, i.e. the size of the AO basis).

The normalised first moment:

$$\bar{\mu}_1 = \mu_1 / \mu_0 = (1/N)\ \Sigma_i\ H_{ii} = \alpha \tag{11}$$

assuming all orbitals are degenerate; this is the baricentre of the MO spectrum.

$$\bar{\mu}_2 = \mu_2/\mu_0 = (1/N)\ \Sigma_{i,j}\ H_{ij}\ H_{ji} = (1/N)\ \Sigma_{i\neq j}\ h_{ij}^2 \tag{12}$$

this is the mean square band width, where, in the nearest neighbour model, $h_{ij} = \beta$ if all bond lengths are equal. Similarly the third and fourth moments are related to the skewness and modality (number of peaks) of the eigenspectrum.

In an interesting example of the application of moments theory, Lee has developed a μ_2-scaled Hückel theory whereby the second moment is used to scale the Hückel energy [23]. Geometries of boranes and carboranes, optimised by this method, have been found to be close to those from Hartree-Fock calculations.

3.3 Graph Theory and Clusters

King and Rouvray [16] presented a graph theoretical treatment of bonding in borane clusters in which the n-atom *closo*-cluster is represented by the complete graph K_n. In the complete graph, all vertices are connected ("adjacent") to all other vertices, by edges of equal weighting (i.e. all off-diagonal elements $H_{ij} = 1$). They showed that the eigenvalue spectrum for a complete graph K_n, with a basis set of n σ-type orbitals, has a single positive eigenvalue (bonding MO) which (in terms of Hückel theory) is given by $E_1 = \alpha + (n\text{-}1)\beta$ and (n-1) degenerate negative eigenvalues (antibonding MOs) $E_2 = E_3 \ldots = E_n = \alpha - \beta$. The com-

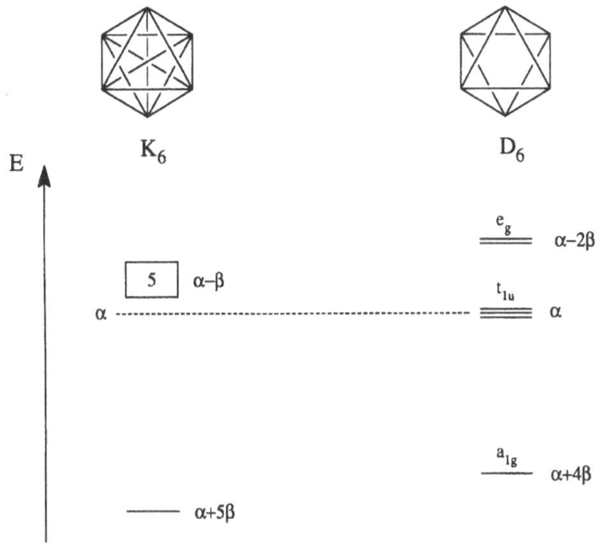

Fig. 3. Eigenvalue spectra (based on the Hückel model) for the completely connected graph K_6 and the deltahedral graph D_6

pletely connected graph K_6 and its eigenvalue spectrum are shown on the left of Fig. 3.

King and Rouvray also showed that, for K_n, the 2n π-symmetry orbitals give rise to n degenerate π-bonding MOs and n degenerate π-antibonding MOs [16]. Wade's (n+1) skeletal electron pair (SEP) rule is rationalised by assuming the occupation of the single bonding σ-MO and the n bonding π-MOs. This method was also applied to the rationalisation of the electron counts of *nido* and *arachno* boranes and capped transition metal clusters [16].

Representing a real cluster by the complete graph K_n is clearly a major oversimplification for $n>4$, since it is impossible in 3-dimensions to have an n-vertex polyhedron ($n>4$) with all $\frac{1}{2}n(n-1)$ "edges" equivalent. It is, of course, possible with $n=2$ (dimer), 3 (triangle) or 4 (tetrahedron). The interactions between n atoms (assuming that the resonance integral β is distance-dependent; $\beta(r)$) can only be equal in a hyper-space of at least (n-1)-dimensions. Thus, for all edges (inter-vertex connections) to be equivalent requires an (n-1)-dimensional polytope.

Since real clusters are 3-dimensional, the spectrum of Hückel σ-type orbitals will not be the same as for the K_n graph. *Closo*-boranes are instead represented by the deltahedral graphs D_n. The "lowering of symmetry" from K_n to D_n ($n>4$) is accompanied by a splitting of the degeneracy of the (n-1) antibonding σ-MOs according to the point group symmetry of the polyhedron. The change in the eigenvalue spectrum on going from K_6 to D_6 (the octahedron) is shown on the right side of Fig. 3.

In 1990, King et al. [24] showed that, in the absence of orbital mixing, the σ-MOs of an octahedral cluster have the following Hückel energies:

$$E_5 = E_6 = E(e_g) = \alpha - (2\text{-}t)\beta_c$$
$$E_2 = E_3 = E_4 = E(t_{1u}) = \alpha - t\beta_c$$
$$E_1 = E(a_{1g}) = \alpha + (4+t)\beta_c$$

where $t = \beta_t / \beta_c$ (the ratio of the *trans* to the *cis* resonance integrals). In the K_6 limit ($t = 1$; $\beta_t = \beta_c = \beta$) the eigenvalue spectrum becomes as in Fig. 3 (i.e. $E_1 = \alpha + 5\beta$; $E_{2..6} = \alpha - \beta$) while in the D_6 limit ($t = 0$; $\beta_t = 0$; $\beta_c = \beta$) the energies are: $E(a_{1g}) = \alpha + 4\beta$; $E(t_{1u}) = \alpha$; $E(e_g) = \alpha - 2\beta$. By fitting to ab initio calculations, King et al. [24] obtained a value of $t = 0.233$, much closer to the D_6 ($t = 0$) than the K_6 ($t = 1$) limit. Similar loss of degeneracy occurs for the π-MOs as the connectivity is reduced from K_n to D_n.

Despite the loss of degeneracy in σ and π-MOs, on going from K_n to D_n, the (n+1)-SEP rule for real deltahedral *closo*-boranes (for which the graph D_n is a good description) can still be accounted for by graph theory since there is still only one strongly bonding σ-MO (Ψ_1) and the non-bonding or weakly anti-bonding σ-MOs (Ψ_2-Ψ_4) are destabilised by mixing with π-type MOs of the same point group symmetry (note that σ/π mixing is forbidden in K_n).

4 Clusters as Spheres

Since in many clusters the atoms (vertices) lie approximately on the surface of a single sphere, a number of attempts have been made to obtain the skeletal MOs of clusters as spherical harmonic expansions by solving the Schrödinger equation for a particle in or on a sphere.

4.1 General Methodology

In spherical cluster models, all of the cluster skeletal atoms are constrained to lie on the surface of a single sphere of radius R, with the location of each atom (i) given by the angular co-ordinates (θ_i, ϕ_i). The Schrödinger equation ($H\Psi = E\Psi$) for a particle on the surface of a sphere becomes

$$-(h^2/2m_eR)\nabla^2\Psi = E_L\Psi_{L,M} \tag{13}$$

where ∇^2 [$= (\sin\theta)^{-1}(\partial/\partial\theta)\cdot\sin\theta(\partial/\partial\theta) + (\sin^2\theta)^{-1}(\partial^2/\partial\phi^2)$] depends only on the angular co-ordinates and the eigenvalues (energies) E_L [$= (h^2/2m_eR)\,L(L+1)$] depend only on the number of angular nodes (L) in the wavefunction. The wavefunctions $\Psi_{L,M}$ are then obtained as LCAO expansions, with the coefficient of the AO (σ_i) on atom i given by the value of the spherical harmonic function $Y_{L,M}(\theta_i, \phi_i)$ at that atom:

$$\Psi_{L,M} = N_{L,M}\,\Sigma_i\,Y_{L,M}(\theta_i, \phi_i)\sigma_i \tag{14}$$

where $N_{L,M}$ is the normalisation constant.

L and M are quantum numbers which are equivalent to the orbital angular momentum (ℓ) and magnetic (m_ℓ) quantum numbers which define atomic orbitals. L takes the values 0 (denoted S by analogy with atomic orbitals), 1(P),

2(D) etc. and, as in the case of atomic orbitals, each L-manifold has a degeneracy of (2L+1) since the M quantum number can take any integer value from -L to +L.

Starting with a basis set of s orbitals (e.g. a hollow cluster of hydrogen or alkali metal atoms), the energy ordering is S < P < D < F < .... The increasing energy with increasing L value reflects the increasing number of nodes in the wave function, since every node is associated with antibonding interactions between the atomic orbitals. While such hollow s-orbital clusters are not known for more than seven atoms, similar ideas form the basis of the "jellium model" of Knight and co-workers [25], which has been used successfully to account for the "magic number" clusters observed (as relatively intense peaks) in the mass spectra of neutral and anionic alkali metal clusters in molecular beams [26]. In the jellium model, the Schrödinger equation is solved for the particle in a sphere problem and the wavefunctions have radial as well as angular parts, in a manner analogous to atomic orbitals themselves [26]. For further details of the jellium model and similar approaches, the reader is referred to the review by Bjørnholm [27].

The first application of spherical harmonic expansions to cluster MOs was by Chapman and Waddington, who obtained the MOs of the cage compounds, S_4E_4 (E=N, As), by solving the Schrödinger equation for a particle on a sphere [28]. The (2L+1)-fold degeneracy of the L manifold was lifted by introducing the actual point group symmetry of the cluster as a perturbation. Hoffmann and Gouterman used a similar analysis for borane clusters, but approximated the perturbation of the spherical harmonics (due to the real cluster geometry) by a Crystal Field expansion [29]. Waddington subsequently adapted his model by constraining the electrons, in boranes and related species, to move along the edges of the cluster polyhedron (i.e. solving the Schrödinger equation for a particle moving along the edges of a polyhedral box) [30].

Despite the appeal of treating clusters as pseudo-spheres, the early treatments obtained energy level orderings which were different from those obtained by conventional LCAO-MO methods, with the exception of trivial cases such

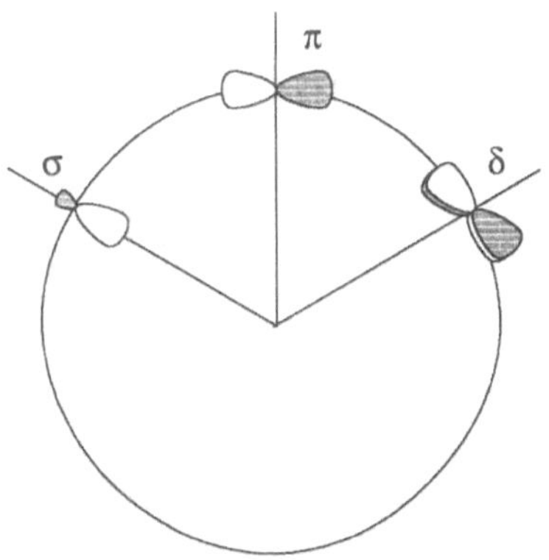

Fig. 4. σ, π and δ-symmetry atomic orbitals on the surface of a cluster sphere

as hollow spherical aggregates of hydrogen atoms (or other s-valent species such as the alkali metals). The reason for this failure is that, in addition to nodeless (σ-type) AOs, most main group and transition metal atoms possess orbitals which have one (π) or two (δ) intrinsic angular nodes in the plane perpendicular to the radial vector, as shown in Fig. 4, which do not contribute to the overall antibonding character of the cluster MOs.

The creation of a more sophisticated spherical cluster model was aided by the work of Hoffman et al., who obtained linear combinations of ligand orbitals in transition metal clusters as projections of Generator Orbitals located at the centre of the complex [31].

4.2
Tensor Surface Harmonic Model

An elegant approach to the bonding in cluster compounds was developed in the early 1980s by A.J. Stone [32–35]. In Stone's Tensor Surface Harmonic (TSH) model the LCAO coefficients are again spherical harmonics but, in contrast to the earlier treatments discussed above, the type of spherical harmonic employed depends on the nodal properties of the AO basis. The TSH method and its applications are discussed in greater detail in other reviews in this volume and elsewhere [1, 8].The treatment below is a brief overview of the method.

For general spherical harmonics ($S_{L,M}$) and AOs (ρ), Eq. (14) becomes:

$$\Psi_{L,M} = N_{L,M} \Sigma_i S_{L,M}(\theta_i, \phi_i)\rho_i \tag{15}$$

For a radial σ-type basis, the required spherical harmonics are the Scalar Spherical Harmonics ($Y_{L,M}$), discussed above, and the cluster MOs are assigned the TSH labels $S^\sigma(L = 0)$, $P^\sigma(L = 1)$, $D^\sigma(L = 2)$ etc., with their energies increasing with L.

For a basis set of tangential π-type AOs, which are singly noded in the plane perpendicular to the radial vector (see Fig. 4), the required harmonics are the "vector surface harmonics". If the cluster is composed of conical E-H or ML_3 fragments, there will be two degenerate mutually orthogonal π-symmetry orbitals at each cluster vertex. These orbitals are denoted π_i^θ and π_i^ϕ as they are oriented along the direction of increasing θ and ϕ respectively. In an n atom cluster, these AOs give rise to 2n tangential or surface cluster MOs.

The vector surface harmonics used in the LCAO expansion are obtained by differentiating the scalar harmonics $Y_{L,M}$ with respect to θ and ϕ. This leads to two sets of vector harmonics for each value of L (except 0 since the S^σ function has no θ or ϕ dependence) which are denoted $V_{L,M}{}^\pi$ and $\bar{V}_{L,M}{}^\pi$:

$$V_{L,M}{}^\pi = \nabla Y_{L,M} \tag{16a}$$

$$\bar{V}_{L,M}{}^\pi = \mathbf{r} \times V_{L,M}{}^\pi \tag{16b}$$

where $\nabla = (\partial/\partial\theta) + (\sin\theta)^{-1}(\partial/\partial\phi)$ and the effect of the operator $\mathbf{r}$ is a local 90° rotation of all the p^π components in $V_{L,M}{}^\pi$ about the radial vector. The cluster MOs are formed as LCAO expansions with the coefficients of the π_i^θ and π_i^ϕ components given by the value the θ and ϕ components of the appropriate vector harmonic at each atom (θ_i, ϕ_i). L can take any integer value (except 0) giving rise to $P^{\pi}/\bar{P}^{\pi}$ (L=1), $D^{\pi}/\bar{D}^{\pi}$ (L=2) etc. sets of MOs. Again each L manifold is (2L+1)-degenerate.

Similar methods are used to generate cluster L^{δ} and $\bar{L}^{\delta}$ orbitals (formed from metal d^{δ} atomic orbitals) – with the second-order "tensor surface harmonics" defined as the concavities of the scalar spherical harmonics:

$$T_{L,M}{}^{\delta} = \nabla\nabla Y_{L,M} \tag{17a}$$

$$\bar{T}_{L,M}{}^{\delta} = \mathbf{r} \times T_{L,M}{}^{\delta} \text{ (45° rotation about the radial vector)} \tag{17b}$$

$T_{L,M}{}^{\delta}$ and $\bar{T}_{L,M}{}^{\delta}$ are second rank tensors whose diagonal and off-diagonal elements give (at the cluster vertices (θ_i, ϕ_i) the values of the coefficients of the two d^{δ} atomic orbitals on each metal) in the LCAO expansion defining L^{δ} and $\bar{L}^{\delta}$ respectively. A detailed study of the L^{δ} and $\bar{L}^{\delta}$ functions of octahedral halide-bridged metal clusters and their *nido*- and *arachno*-derivatives enabled Johnston and Mingos to rationalise the electron counts of square and pyramidal molybdenum clusters [36].

For the particle on a sphere problem the L^{π} orbitals are all strictly bonding, becoming less strongly bonding as L^{π} rises. The $\bar{L}^{\pi}$ orbitals, on the other hand, are all strictly antibonding, becoming less anti-bonding as L increases. This mirror image behaviour of L^{π} and $\bar{L}^{\pi}$ is quite general and is due to the fact that they can be interconverted by a local rotation of each component π-type AO by 90° about the radial vector, leading to the inter-conversion of bonding and antibonding character [32].

Stone [32–35] used TSH theory to rationalise Wade's (n+1) SEP rule for deltahedral clusters, such as the *closo*-boranes (closest to the idealised spherical limit) by recognising that the unique S^{σ} orbital will be bonding, as will all n L^{π} MOs, with the remaining (n-1) L^{σ}MOs destabilised due to mixing with symmetry matched (and lower lying) L^{π} MOs. MNDO calculations by Brint et al. [37] confirmed that the qualitative TSH approach leads to a good description of the bonding in deltahedral boranes.

For an n-vertex cluster, there are n unique L^{σ} functions and n $L^{\pi}/\bar{L}^{\pi}$ pairs and, in general, the first n (i.e. the lowest L values) are utilised, provided that:

(i) they do not possess zero amplitude at all cluster vertices ("null functions") and
(ii) they are not repeats of earlier MOs ("redundant functions").

For a truly spherical cluster (corresponding to an infinite number of vertices distributed over the cluster sphere) there would be no null or obsolete functions, so the occurrence of such deviations is a consequence of the finite and non-spherical nature of real clusters. The valid L^{σ} (and higher order harmonic) functions and the occurrence of gaps in L manifolds have been enumerated by Johnston and Mingos for a variety of polyhedra [38].

The finite and non-spherical nature of clusters also leads to loss of the (2L+1)-fold degeneracy of a given L manifold. This problem, together with the occurrence of σ-π mixing was addressed by Fowler and Porterfield in their Extended Tensor Surface Harmonic method [39], where the cluster energies were evaluated within the Extended Hückel framework, taking into account the actual point group symmetry of the cluster.

In conclusion, the importance, and usefulness of TSH theory has been summed up succinctly by Fowler and Porterfield [39]: "*the strength of TSH theory is that its pseudo-spherical symmetry labels provide a language for the qualitative discussion of cluster bonding that is often useful in interpreting more quantitative calculations*".

5
Group Theory of Clusters

So far in this review I have made use of group theoretical terms, both from finite point groups (such as O_h and I_h) and, in the case of TSH theory, the infinite spherical group (R_3), to label cluster MOs. These labels are extremely useful and permit the enumeration of symmetry-allowed spectroscopic transitions for cluster molecules and also the derivation of symmetry-based rules concerning the allowed or forbidden nature of polyhedral rearrangements and fluxional processes [40–43].

In addition to these applications, the past decade has seen the development of a number of elegant rules for obtaining the symmetries of cluster MOs (as opposed to the de facto assignment of symmetry by inspection of the orbitals) within the TSH framework. Group theory has also been used to explain deviations from Wade's (n+1)-SEP rule in a number of cases.

5.1
Group Theory on a Spherical Shell

In a series of seminal papers, Quinn and co-workers developed a group theoretical formalism in which symmetry adapted linear combinations of ligand orbitals are generated within the TSH framework [44–46]. This method has been shown to be equally applicable to cluster compounds, enabling the derivation of the valid L^σ, $L^\pi/\bar{L}^\pi$ and $L^\delta/\bar{L}^\delta$ orbitals by considering point group symmetries on the spherical shell. Fowler [47, 48] and Johnston and Mingos [49, 50] subsequently applied Quinn's methodology to the study of a wide range of cluster molecules. Ceulemans developed a related approach using induction theory to link the local site symmetry of the cluster vertices and the symmetries spanned by the cluster orbitals [51]. For a detailed discussion of Quinn's group theoretical methodology and extensions of the approach, see the review by Johnston and Mingos [1].

The first step in Quinn's approach is to determine the point group (G) of the cluster polyhedron [52]. The next stage is to obtain the "permutation representation" of the cluster polyhedron. The permutation representation (Γ_ρ) is a reducible representation such that the character (χ_ρ(R)) for each operation (R) of the point group is equal to the number of cluster vertices which remain fixed under that operation [52]. Quinn showed that the reducible representations (Γ_ℓ) corresponding to the set of linear combinations of s ($\ell=0$), p ($\ell=1$), d ($\ell=2$) etc. atomic orbitals, located at the cluster vertices, can be obtained by taking the product of the permutation representation with the symmetry of the appropriate spherical harmonic ($Y_{\ell,m}$) at the centre of the cluster (Γ_ℓ^0):

$$\Gamma_\ell = \Gamma_\rho \otimes \Gamma_\ell^0 \quad (18)$$

The representation Γ_ℓ is then reduced (obtained as a sum of irreducible representations) in the standard manner [52].

For $\ell = 0$, and because of the nodeless nature of s orbitals, Γ_s^0 is the totally symmetric representation Γ_g^0 (with $\chi(R) = +1$ for all operations). Since multiplication by Γ_g^0 leaves any representation unchanged, Eq. (18) simplifies to $\Gamma_s = \Gamma_\rho$. A single s orbital at each cluster vertex will generate the cluster L^σ orbitals and the symmetries of the L^σ orbitals of any cluster can simply be obtained as

$$\Gamma_\sigma = \Gamma_\rho \quad (19)$$

For $\ell = 1$, since the three p orbitals at a cluster vertex can be combined to yield one radial (p^σ) and two tangential (p^π) orbitals, then

$$\Gamma_p = \Gamma_\rho \otimes \Gamma_p^0 = \Gamma_\sigma \oplus \Gamma_\pi \quad (20)$$

so that the symmetries spanned by the L^π and $\bar{L}^\pi$ orbitals are given by

$$\Gamma_\pi = \Gamma_\sigma \otimes \Gamma_{x,y,z} - \Gamma_\sigma \quad (21)$$

since Γ_p^0 is the representation spanned by the x, y and z vectors at the cluster centre.

Similarly, for the L^δ and $\bar{L}^\delta$ orbitals:

$$\Gamma_\delta = \Gamma_\rho \otimes \Gamma_d^0 - \Gamma_\sigma - \Gamma_\pi \quad (22)$$

Thus, knowing the permutational symmetry of the cluster and how the generator functions Γ_s^0, Γ_p^0, Γ_d^0 etc. transform under the point group symmetry of the cluster, enables the symmetry of the cluster σ, π, δ etc. orbitals to be determined. Finally, it should be noted that the symmetries of the L^π and $\bar{L}^\pi$ orbitals (and the $L^\delta/\bar{L}^\delta$ MOs) are interconverted by multiplying by the representation (denoted Γ_u^0) which has $\chi(R) = +1$ for all proper rotations and $\chi(R) = -1$ for all improper rotations [1, 44–46].

5.2 Group Theory of Orbits

The methodology of group theory on a spherical shell was extended in an elegant and instructive way by Fowler and Quinn [53] who introduced the concepts of orbits. Within a given point group G there are a number of realisable orbits (polyhedra where all vertices are equivalent) of the point group, depending on where the vertices are positioned. For example, in the octahedral point group (O_h) one orbit is defined by the vertices that lie on the 4-fold rotation axes. These vertices have local site-symmetry C_{4v} and the orbit has size 6 (the 6 vertices define an octahedron). The number of equivalent vertices in an orbit is given by

$$N_O = h_G / h_S \quad (23)$$

where h_G is the order of the cluster point group G (48 in the O_h point group) and h_S is the order of the vertex point group (i.e. the site symmetry S) (8 for C_{4v}). The O_6 orbit of the point group O_h is therefore the octahedron. Similarly the O_8 orbit corresponds to the cube.

Fowler and Quinn tabulated the symmetries of the σ and π cluster MOs for all the orbits of all high symmetry and many cyclic and dihedral point groups [53]. These tables can be used to determine the representations spanned by the σ and π MOs of any polyhedral cluster as follows:

(i) divide the polyhedron up into its constituent orbits (sets of symmetry-equivalent vertices) O_i, O_j, O_k ...
(ii) obtain the total set of cluster MO symmetries by summing those of the constituent orbits – e.g. $\Gamma_\pi = \Gamma_\pi(O_i) \oplus \Gamma_\pi(O_j) \oplus \Gamma_\pi(O_k) \oplus ....$

This decomposition into orbits has been particularly important in rationalising deviations from Wade's rules in polar deltahedra (see below) and in studying the fullerenes.

6
Cluster Topology: Localised Orbital Treatments

The application of localised bonding schemes to clusters is probably the oldest theoretical approach to cluster bonding. In such models, electron pairs are associated with certain topological features of the cluster polyhedron (generally the edges or faces). Although such models may be regarded as applications of graph theory, they also depend on the 3-D arrangement of the cluster vertices in space – i.e. the polyhedral geometry. As will become apparent, many of the concepts and methods introduced in Sects. 2–5 have been adopted in localised bonding treatments.

6.1
Valence Bond Models

The first clusters to be studied from a localised Valence Bond (VB) [54] viewpoint were the boron hydride cage compounds, B_nH_{n+m}. The VB approach describes bonding in terms of electron pair bonds between neighbouring atoms.

The "Effective Atomic Number" (EAN) rule [8], adhered to by "electron-precise" clusters such as tetrahedral P_4 and the trivalent hydrocarbon clusters, C_nH_n, is based on VB theory. In the EAN model, it is assumed that each edge of the cluster polyhedron corresponds to a bond pair. Since edges, and hence bond pairs, are shared by two vertices (atoms) and because of the tendency of main group and transition metal atoms to achieve a closed shell (inert gas) configuration, the number of valence electrons (N_e) of electron-precise clusters is given by

$$N_e\,(\text{main group}) = 8V - 2E \qquad (24a)$$

$$N_e\,(\text{transition metal}) = 18V - 2E \qquad (24b)$$

The EAN rule, however, cannot be simply applied to the polyhedral boranes, which possess more edges than bond pairs. In 1949, Longuet-Higgins stated that any localised bonding description of the bonding in such molecules must include 3-centre-2-electron (3c-2e) bonds in addition to classical 2-centre-2-electron (2c-2e) bonds [55].

Localised bonding schemes involving 2-centre and 3-centre bonds (BH, BHB, BB and BBB) were subsequently developed by Duffey [56] and Lipscomb et al. [57] who showed that the hypothetical tetrahedral species B_4H_4 (iso-electronic with B_4Cl_4) has four bond-pairs located at the centres of the four triangular faces.

6.2 The styx Model

In the 1960s, Lipscomb developed a topological methodology, involving 2c-2e and 3c-2e bonds, which proved useful in rationalising the structures of low-symmetry boranes of the form B_nH_{n+m} [58]. The *styx* formalism starts from the premise that each boron atom may participate in four types of bond: B-H-B (3c-2e); B-B-B (3c-2e); B-B (2c-2e); B-H (2c-2e). For a general borane, B_nH_{n+m}, with n *exo*-terminal BH units and m *endo*-terminal or bridging hydrides, the total number of bond pairs is given by

$$N = s + t + y + x = 2n + (m/2) \qquad (25)$$

(since each boron atom contributes three valence electrons and each hydrogen one electron) where s, t, y and x are the numbers of B-H-B, B-B-B, B-H and B-B bonds respectively. By counting atoms, orbitals and electrons, Lipscomb derived the following relationships:

$$x = m - s; \quad t = n - s; \quad 2y = s - x \qquad (26)$$

and assigned each cluster a set of styx indices. This method successfully accounts for the geometries and charge distributions in low symmetry borane cage compounds but for the higher symmetry boranes (in particular the anions $[B_nH_n]^{2-}$) the *styx* formalism leads to an unworkable number of resonance forms (required to maintain high symmetry) because the bonding in such clusters can no longer adequately be described within a simple localised framework [59].

6.3 The Topological Equivalent Orbital Model

6.3.1 *Methodology*

The relationship between delocalised and localised cluster bonding descriptions can be appreciated within an Equivalent Orbital framework. A basis set of atomic orbitals (AOs) can be transformed into an alternative, but equivalent, basis set of hybrid orbitals which are known as "equivalent orbitals" (EOs).

Taking linear combinations of these EOs generates MOs which have identical nodal and symmetry properties to LCAOs formed from the original AO basis set [60].

In the 1960s, Kettle and Tomlinson [61] developed the Topological Equivalent Orbital (TEO) method for describing the bonding in main group and transition metal clusters. The TEO approach uses basis sets of edge- or face-localised ("topological") equivalent bonding orbitals. Linear combinations of localised orbitals (LCLOs) are generated which have identical symmetry properties and similar energies to those obtained from more sophisticated LCAO-MO calculations. The TEO method differs from previous localised orbital treatments because the edge- or face-localised orbitals are allowed to interact (i.e. in Hückel terms there is a non-zero off-diagonal Hamiltonian matrix element or resonance integral between the localised orbitals). Within the Hückel MO framework, each localised orbital is assigned a Coulomb energy (α_f for face-localised and α_e for edge-localised) and all resonance integrals are set to zero except those between faces which share a common edge (β_f) and between edges which are coincident and which bound a common face (β_e).

6.3.2
Borane Clusters

The results of Kettle and Tomlinson's calculations [61b] for octahedral $[B_6H_6]^{2-}$, together with those of an LCAO study by Longuet-Higgins and Roberts [62] are reproduced in Table 2.

Kettle and Tomlinson proposed that the linear combinations of edge-localised orbitals with energies $E_e > \alpha_e - 0.8(\pm 0.2)\beta_e$ and face-localised orbitals with energies $E_f > \alpha_f - 1.8(\pm 0.2)\beta_f$ correspond to antibonding (and hence unoccupied) rather than bonding skeletal MOs [61]. For $[B_6H_6]^{2-}$ there are seven linear combinations (with symmetries a_{1g}, t_{1u} and t_{2g}) which have energies below these limits (using either the edge-localised or the face-localised basis) and which therefore correspond to the seven skeletal bonding MOs of the octahedral cluster.

The success of Kettle and Tomlinson's EO approach indicates that the resonance integrals between the localised orbitals are significant and that the Hückel

Table. 2. Hückel energies of localised orbitals for $[B_6H_6]^{2-}$ (adapted from [61b])

Description	Symmetry	Edge basis (E_e)	Face basis (E_f)	Reference [62]
Unoccupied MOs[a]	α_{2u}		$\alpha_f - 3\beta_f$	
	e_g	$\alpha_e - 2\beta_e$		-0.88χ[b]
	t_{2u}	$\alpha_e - 2\beta_e$		-0.42χ
Occupied MOs	t_{2g}	α_e	$\alpha_f - \beta_f$	0.49χ
	t_{1u}	$\alpha_e + 2\beta_e$	$\alpha_f + \beta_f$	1.02χ
	a_{1g}	$\alpha_e + 4\beta_e$	$\alpha_f + 3\beta_f$	2.97χ

[a] LCLO functions equivalent to LCAO functions based on atomic d^δ orbitals.
[b] The parameter χ is an approximately linear function of energy.

approximation (which introduces the appropriate nodal surfaces into the LCLO functions) provides a satisfactory method of establishing which linear combinations are low-lying. Kettle and Tomlinson demonstrated [61b] that the TEO approach represents a topologically correct extension of Hückel theory to three dimensions because of the isomorphism which exists between the Hückel Hamiltonian matrix and topological matrices relating to the edges and faces of the cluster polyhedron. This is entirely equivalent, in the 3-D space of the cluster molecule, to the relationship between Hückel and graph theory discussed above.

6.3.3
Transition Metal Clusters

In a VB treatment of octahedral $[Mo_6Cl_8]^{4+}$, which has 8 face-bridging chloride ligands, Sheldon [63] and Duffey et al. [64] concluded that the cluster has 12 2c-2e bonds which are localised on the 12 edges of the octahedron. Kettle [65] subsequently confirmed that the 12 skeletal bonding MOs (from the MO calculations of Cotton and Haas [66]) are topologically equivalent to LCLOs based on the set of 12 σ-type edge-bonding (2c-2e) orbitals. He also showed that octahedral clusters of the type $[M_6Cl_{12}]^{2+}$ (M = Nb, Ta), with 12 edge-bridging chloride ligands, possess 8 cluster bonding orbitals which are topologically equivalent to linear combinations of the 8 face-bonding (3c-2e) localised orbitals [65].

Kettle also demonstrated that tetrahedral clusters of the main group and transition metal elements are characterised by six edge-localised bonding orbitals [68]. Thus $[Mo_6Cl_8]^{4+}$ (O_h) and P_4(T_d) are both examples of "electron-precise" clusters because there is a one-to-one topological mapping between the edges of the cluster polyhedron and the cluster skeletal bonding MOs.

Turning to metal carbonyl clusters, Kettle used the TEO method to study the bonding in the octahedral cluster $[Rh_6(CO)_{16}]$ and came to the conclusion that, as in the *closo*-boranes, this molecule is electron deficient, as there are insufficient electrons available to form 12 2c-2e localised bonds [68]. He was not, however, able to separate properly the metal-metal and metal-carbonyl bonding orbitals, so that a direct comparison with the boranes was impossible. In fact, the discovery of the link between the bonding in borane and metal carbonyl clusters had to await the more detailed MO calculations of Mingos [7] and Lauher [69].

6.4
The Tensor Surface Harmonic-Equivalent Orbital (TSH-EO) Model

Since the centroids of the faces, or the midpoints of the edges, of many clusters lie approximately on a single sphere, the methodology of TSH theory can be invoked to express the LCLO functions as spherical harmonic expansions. In 1989, Johnston, et al. [70] adapted and extended Kettle's TEO approach within the framework of TSH theory.

In the TSH-EO model, a basis set of localised orbitals are defined at the midpoints of the cluster edges (edge-localised orbitals; ELOs; λ_e) or the centroids of

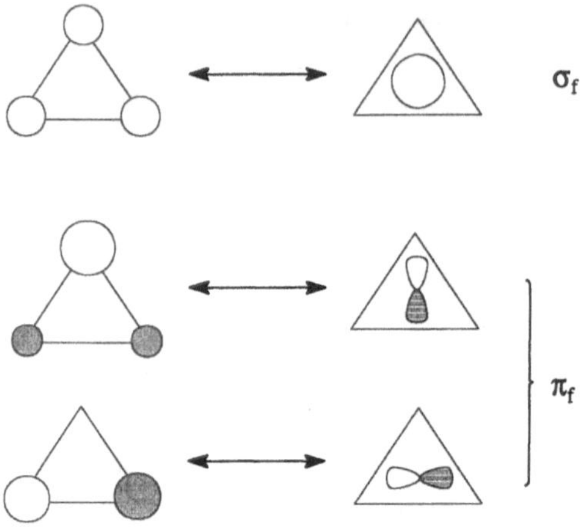

Fig. 5. Examples of equivalent LCAO and face-localised orbitals

the cluster faces (face-localised orbitals; FLOs; λ_f). These localised orbitals (LOs) are assigned axial symmetry labels (σ, π, δ etc.), as in TSH theory, depending on their nodal properties.

Each cluster edge has two LOs associated with it (σ_e and π_e) which are equivalent to bonding and antibonding combinations of atomic orbitals which are oriented along the cluster edge. The number of FLOs associated with a given face is equal to the size of the face. Thus a triangular face has three FLOs ($\sigma_f + 2\pi_f$, as shown in Fig. 5) and a square face has four FLOs ($\sigma_f + 2\pi_f + \delta_f$). Within the Hückel approximation, the energies of these FLOs increase with the number of angular nodes (corresponding to antibonding interactions) that they possess, though the FLOs with a given number of angular nodes become more stable as the size of the cluster face increases.

For localised orbitals, Eq. (15) becomes:

$$\Psi_{L,M} = N_{L,M}\ \Sigma_i\ S_{L,M}(\theta_i, \phi_i)\lambda_i \tag{28}$$

where θ_i and ϕ_i are the loci of the edge- or face-localised EOs (λ_i). Following the nomenclature of TSH theory, the LCLOs are denoted $L^{e\sigma}$, $L^{f\sigma}$ $L^{e\pi}/\bar{L}^{e\pi}$ etc. The application of the TSH-EO model to hydrocarbon clusters will be discussed in Sect. 7.2.

7 Case Studies

The field of theoretical cluster chemistry (of which mathematical cluster chemistry is a sub-discipline) was inspired by the early synthesis and structural characterisation of the *closo*-, *nido*- and *arachno*-boranes and the subsequent development of metal cluster chemistry. This eventually led to the generation of a number of general rules which are applicable to a wide range of clusters [1, 4–7]. A number of deviations from these rules have, however, become apparent and these, along with the characterisation of completely new cluster structure types, have given new life and vigour to mathematical cluster chemistry. Rationalisation of these exceptions has led mathematical cluster chemists to utilise a variety of the techniques described above and has led to new rules and generalisations. Some examples are presented here as case studies.

7.1 Deviations from Wade's Rules: "Non-Spherical" Clusters

It was shown in Sect. 4 that TSH theory explains Wade's (n+1) SEP rule for deltahedral boranes [4–6]. Deviations from Wade's rule are thus associated with a breakdown of the premises of TSH theory and occur when a cluster polyhedron is not sufficiently pseudo-spherical for the particle on a sphere approximation to be a reasonable approximation. A combination of semi-empirical MO calculations and group theoretical analysis have enabled such deviations to be rationalised within the TSH framework.

Stone showed that pyramidal *nido*-boranes, which have (n+2) rather than (n+1) SEPs, are characterised by a pair of non-bonding orbitals constituting a degenerate $L^{\pi}/\bar{L}^{\pi}$ pair [33]. Because of the mirror-image bonding characteristics of L^{π} and $\bar{L}^{\pi}$ MOs, Stone demonstrated that $L^{\pi}/\bar{L}^{\pi}$ degeneracy is only possible if the orbitals are non-bonding. The same argument applies to tetrahedral clusters such as P_4 and C_4H_4 which, although having all triangular faces, may be regarded as *nido*-trigonal bipyramids, with six rather than five SEPs and a degenerate (e-symmetry) $L^{\pi}/\bar{L}^{\pi}$ pair of HOMOs [33].

Fowler extended this idea as "The Pairing Principle" [47] and invoked group theoretical arguments (based on Quinn's methodology – as outlined in Sect. 5) to explain departures from the (n+1) SEP rule predicted by Lipscomb, on the basis of MO calculations [71], for the hypothetical high nuclearity deltahedral boranes $[B_{16}H_{16}]$ (T_d), $[B_{19}H_{19}]$ (C_{3v}) and $[B_{22}H_{22}]$ (T_d). Fowler deduced that, if the representation spanned by the cluster $L^{\pi}/\bar{L}^{\pi}$ orbitals (under the point group symmetry of the cluster) contains an odd number of degenerate e-type representations then, because of the Pairing Principle, one of the e-pairs must be a degenerate non-bonding $L^{\pi}/\bar{L}^{\pi}$ pair. Fowler [47] and Johnston and Mingos [72] showed that for deltahedra this situation arises when there is a single atom on each C_3 rotation axis of a deltahedron belonging to the point groups C_3, C_{3v}, T or T_d. If there is a single atom on a 4- or 5-fold axis then this leads to *nido*-polyhedra with open square or pentagonal faces respectively and a degenerate frontier $L^{\pi}/\bar{L}^{\pi}$ pair which can be stabilised by bridging hydrides to give an (n+2) SEP count.

Johnston et al. also showed [72–74] that the Pairing Principle can be used to explain the occurrence of "*hyper-closo-*"metallaboranes and metallacarboranes with n, rather than (n+1) SEPs, and again one atom on the principal rotation axis. The term "polar deltahedra" was introduced to describe such polyhedra [72]. This work was extended to enable a classification of deltahedra [49] in terms of their topological features (e.g. the number of atoms lying on the principal axes) and to rationalise occurrence and nature of deviations from the (n+1) SEP rule. The extension of the TSH theory to include bispherical clusters (where the skeletal atoms lie on the surfaces of two different spheres) was also introduced, thus enabling the study of capped clusters [50].

7.2
Hydrocarbon Cages: Grossly "Non-Spherical" Clusters

The synthesis, in the 1970s, of a series of hydrocarbon molecules, C_nH_n, with trivalent polyhedral structures (see Fig. 1), led to a revival in interest in localised bonding schemes. By following an elegant and complex synthetic pathway, the "holy grail" of hydrocarbon cages – dodecahedral $C_{20}H_{20}$ (with I_h symmetry) – was eventually reported by Paquette et al. in 1982 [75]. The carbon skeleton of dodecahedrane is shown in Fig. 6.

As Kettle showed for tetrahedral clusters [67], the skeletal bonding orbitals in the "electron-precise" molecules C_nH_n are topologically equivalent to linear combinations of edge-localised 2c-2e bond pairs. Schulman and co-workers performed ab initio calculations on cubane, C_8H_8, utilising a basis set of 2c-2e localised C-C and C-H bonding orbitals [76]. The resonance integrals between C-C localised orbitals were large, leading to a broad spectrum of skeletal MOs. Schulman's methodology was similar to that of Kettle, but he also included interaction between cage (C-C) and ligand (C-H) bonding orbitals, leading to an energy level ordering identical to that obtained by conventional LCAO-MO calculations [77]. A similar EO method was applied by Heilbronner et al. in a study of tetrahedral $C_4Bu^t_4$ [78]. Recent 3-D Hückel calculations by Gimarc and Zhao confirm the localised nature of the bonding in these molecules [20a,c].

In 1985, Johnston and Mingos reported the application of a combined extended Hückel-TSH study of trivalent hydrocarbon clusters [79]. They found that the spectrum of cluster MOs was such that the 3n/2 occupied skeletal MOs can be split into a strongly bonding set of (n/2+2) orbitals and a set of (n-2) orbitals which are only weakly bonding, or approximately "non-bonding". The calculated energies and TSH and group theoretical labels of the frontier orbitals of a number of C_nH_n clusters are listed in Table 3 [80], from which it is apparent that the (n-2) "non-bonding" orbitals are made up of (n/2–1) L^π MOs and their (n/2 – 1) $\bar{L}^\pi$ counterparts. This is another example of the pairing principle, since if an L^π MO is weakly bonding/non-bonding then its $\bar{L}^\pi$ counterpart must be weakly antibonding/non-bonding. The splitting of the occupied skeletal MOs, postulated on the basis of extended Hückel calculations, is confirmed by

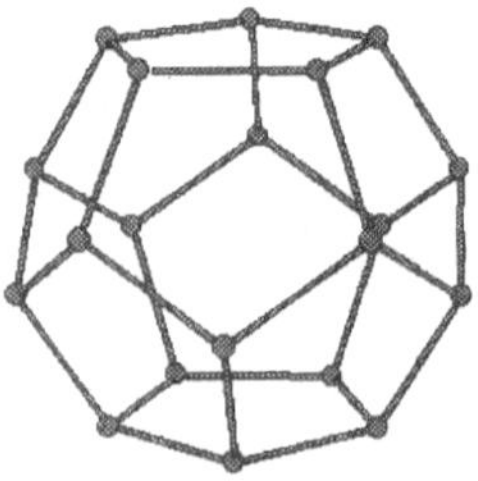

Fig. 6. The carbon skeleton of dodecahedrane, $C_{20}H_{20}$

Table. 3. Extended Hückel energies of the frontier orbitals of some trivalent hydrocarbon clusters, C_nH_n [80]. The "non-bonding" orbitals are in bold. Energies should be compared with the C (2p) AOs (energy = −11.4 eV). Idealised geometries (C-C = 1.54 Å; C-H = 1.09 Å) were adopted

C_4H_6-tetrahedron (T_d)

Orbitals	Energy (eV)
$\bar{P}^\pi$ (t_1)	- 4.33
D^π/$\bar{D}^\pi$ (e)	**- 11.95**
P^+ (t_2)	- 14.33[a]
S^σ (a_1)	- 29.54

C_6H_6-trigonal prism (D_{3h})

Orbitals	Energy (eV)
$\bar{D}^\pi$ (a_1'')	- 1.95
$\bar{P}^\pi$ (a_2'')	- 6.15
D^π (e″)	**- 11.74**
$\bar{D}^\pi$ (e′)	**- 12.93**
P^+ (e′)	- 14.74
P^+ (a_2'')	- 14.77
D^π (a_1')	- 14.78[b]

C_8H_8-cube (O_h)

Orbitals	Energy (eV)
$\bar{D}^\pi$ (e_u)	+ 2.79
$\bar{P}^\pi$ (t_{1g})	- 1.67
D^π (t_{2g})	**- 12.05**
$\bar{D}^\pi$ (t_{2u})	**- 12.61**
D^π (e_g)	- 15.10
P^+ (t_{1u})	- 15.14

$C_{10}H_{10}$-pentagonal prism (D_{5h})

Orbitals	Energy (eV)
$\bar{P}^\pi$ (e1″)	- 2.10
$\bar{F}^\pi$ (e2′)	**- 11.79**
F^π (e2″)	**- 12.50**
D^π (e1″)	**- 12.88**
$\bar{D}^\pi$ (e1′)	**- 13.12**
D^π (e2″)	- 14.81
D^π (a1′)	- 15.13[c]

$C_{12}H_{12}$-hexagonal prism (D_{6h})

Orbitals	Energy (eV)
$\bar{P}^\pi$ (e_{1g})	- 1.52
F^π (b_{2u})	**- 11.12**
$\bar{F}^\pi$ (e_{2g})	**- 11.89**
F^π (e_{2u})	**- 12.25**
$\bar{F}^\pi$ (b_{2g})	**- 12.94**
D^π (e_{1g})	**- 13.59**
$\bar{D}^\pi$ (e_{1u})	**- 13.68**
D^π (e_{2g})	- 14.66
D^π (a_{1g})	- 15.07
F^π (b_{1u})	- 15.25

$C_{12}H_{12}$-truncated tetrahedron (T_d)

Orbitals	Energy (eV)
$\bar{P}^\pi$ (t_1)	- 4.58
$\bar{D}^\pi$ (e)	**- 11.80**
F^π (t_1)	**- 12.70**
$\bar{F}^\pi$ (t_2)	**- 13.78**
D^π (e)	**- 13.99**
D^π (t_2)	- 14.25
F^π (a_1)	- 14.93

$C_{20}H_{20}$-dodecahedron (I_h)

Orbitals	Energy (eV)
$\bar{P}^\pi$ (t_{1g})	- 1.36
$\bar{F}^\pi$ (g_g)	**- 12.64**
$\bar{G}^\pi$ (h_u)	**- 12.88**
G^π (h_g)	**- 13.28**
F^π (g_u)	**- 13.28**
F^π (t_{2u})	- 14.35
D^π (h_g)	- 15.33

[a] The lowest P orbitals are an in-phase combination of P^σ and P^π.

[b] The low-lying S orbital has been omitted for $n > 4$.

[c] The P^+ orbitals have been omitted for $n > 8$.

comparison with more sophisticated semi-empirical and ab initio calculations [77, 81, 82] and also, in the case of cubane, by photoelectron spectroscopy.

For the sub-set of prismatic trivalent clusters, it was shown that the non-bonding character of these MOs arose from an approximate cancellation of bonding and anti-bonding interactions within and between the parallel rings of which the prism is composed [79]. Trivalent polyhedra can be described as "grossly non-spherical" because, although their vertices may lie on the surface of a single sphere, they do not represent a sufficiently good coverage of the sphere to cause all of the L^{π} MOs to be bonding (as in the deltahedral boranes) and the $\bar{L}^{\pi}$ MOs to be antibonding. Johnston and Mingos have also shown, however, that triangular faces are not a prerequisite for "spherical behaviour" since tetravalent polyhedra (having some square or rhombic faces) are also predicted to have (n+1) SEPs [83].

Johnston et al. subsequently rationalised the spectra of skeletal MOs for all trivalent clusters using localised bonding arguments, within the TSH-EO framework [70, 80]. It was noted that the number of "non-bonding" orbitals (n-2) is, by rearranging Euler's equation (1) equal to E-F (i.e. the number of edges of the polyhedron minus the number of faces). Since it is known that trivalent clusters have 3n/2 occupied skeletal MOs and that this is equal to the number of edges of the polyhedron (Sect. 2.2), Johnston et al. realised that the strongly bonding orbitals correspond to the sub-set of linear combinations of (σ-type) edge-bonding orbitals which are also bonding with respect to the faces (F = n/2 + 2 for trivalent polyhedra) of the polyhedron. The remaining E-F (= n − 2) occupied skeletal MOs therefore correspond to linear combinations of edge-bonding orbitals which are noded on some of the cluster faces (i.e. are face non- or anti-bonding) and are therefore higher in energy. A schematic spectrum of skeletal MOs for trivalent clusters is shown in Fig. 7.

This localised picture has been verified, in all the cases studied, by considering the symmetries of the skeletal MOs (see Table 3) since

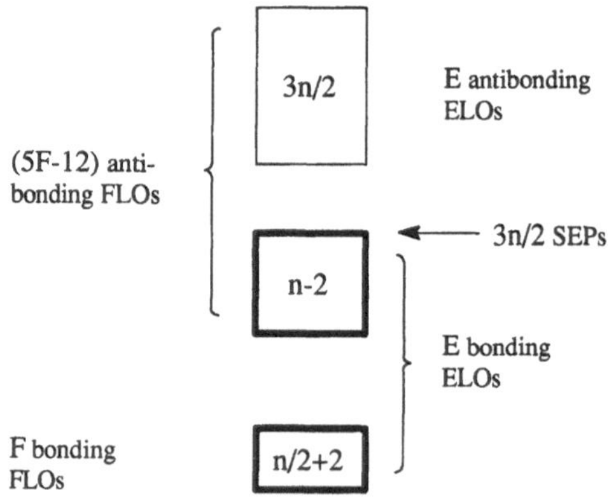

Fig. 7. Schematic skeletal MO diagram for trivalent clusters. n, E and F are the numbers of atoms (= vertices, V), edges and faces of the cluster. The 3n/2 SEPs correspond to the blocks of MOs *outlined in bold*

$$\Gamma_{occ}(P) = \Gamma^{\sigma}(E) \quad (29a)$$
$$\Gamma_{bond}(P) = \Gamma^{\sigma}(D) \quad (29b)$$
$$\Gamma_{non\text{-}bond}(P) = \Gamma^{\sigma}(E) - \Gamma^{\sigma}(D) \quad (29c)$$

where E and D represent the edge-figure and (face-)dual polyhedra of the trivalent polyhedron P.

For the deltahedral boranes, which have too many edges and (for $n>5$) too many faces to be electron precise, there are a number of linear combinations of edge- and face-localised orbitals which are "forbidden", in that they possess too many nodes to be equivalent to linear combinations of AOs with σ and π character only [70]. The remaining ("allowed") LCLO functions, it transpires, are both edge- and face-bonding, which explains why deltahedral clusters (with the exception of the polar deltahedra) have a set of $(n+1)$ bonding skeletal MOs which are well separated from the antibonding MOs, and no non-bonding orbitals [70].

For a comprehensive review of rules for predicting non-bonding MOs in cluster and coordination compounds see Mingos and Lin [84].

7.3 The Fullerenes: a Mathematical Chemist's Delight

The past decade has seen an explosion of interest in the cluster chemistry of carbon, following the seminal work of Kroto et al. who (in 1985) detected an intense feature in the mass spectrum of the products of the laser vaporisation of graphite and postulated that the peak was due to the highly symmetrical truncated icosahedral C_{60} cluster "buckminsterfullerene", shown in Fig. 8 [85, 86]. Calculations, of varying degrees of sophistication, indicated the stability of such a species and, in 1990, Huffman et al. succeeded in separating and crystallising macroscopic quantities of C_{60} and the geometric structure was confirmed [87]. The spectroscopy and chemistry of buckminsterfullerene and a whole series of analogous clusters (named the "fullerenes") has been discussed in detail in a number of reviews and books [88, 89].

The "discovery" of the fullerenes has also led to a renaissance of interest in the application of mathematics to cluster molecules, partly due to the aesthet-

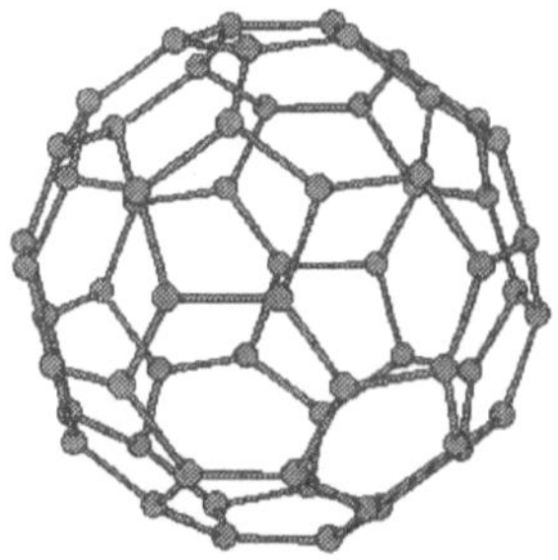

Fig. 8. Buckminsterfullerene, C_{60}

ically pleasing highly symmetric nature of buckminsterfullerene (C_{60}), but also because enumerating the number of possible fullerene isomers for a given nuclearity is a vast combinatorial problem and, in order to predict the most stable isomer (the global minimum on the potential energy hypersurface), rules had to be found to relate the electronic and geometric structures of the various isomers. To this end, the whole battery of mathematical techniques discussed in this review have been brought to bear on the fullerenes, often producing results which have proved of considerable use in the study of other types of cluster molecules. More detailed reviews of work in this field have been presented by Fowler and Manolopoulos [11, 12, 90] and Trinajstic et al. [19d].

7.3.1 *Fullerene Polyhedra*

The fullerenes are trivalent polyhedral clusters with pentagonal and hexagonal faces only. In a VB sense, the cluster framework is formed by overlap of carbon sp^2 hybrids, so that a C_n cluster has 3n/2 bonding and 3n/2 antibonding framework orbitals. The remaining n p-orbitals (one on each carbon atom) are oriented radially and would be utilised in bonding to hydrogen in a hydrocarbon cluster. In terms of TSH theory, these orbitals give rise to n radial L^σ combinations, though they are often described as π-orbitals [11, 12] since the interaction between neighbouring radial p-orbitals on the cluster surface has predominantly π character, especially for large clusters. In the following treatment I shall use the TSH label L_p^σ to refer to these orbitals.

For trivalent polyhedra, Eq. (2a) for the number (f_k) of faces of a given size (k) reduces to [13]

$$\Sigma_k f_k (6\text{-}k) = 12 \tag{30a}$$

or

$$3f_3 + 2f_4 + f_5 + 0f_6 - f_7 - 2f_8 \ldots = 12 \tag{30b}$$

For fullerenes (with only pentagonal and hexagonal faces) Eq. (30b) reduces further to $f_5 = 12$, which explains why all fullerenes have 12 pentagonal faces. The number of hexagonal faces (f_6) is undefined as it has zero weighting in Eq. (30) and is therefore a variable which gives rise to the family of fullerenes. It is not, however, possible to construct a closed trivalent polyhedron with 12 pentagonal and one hexagonal face [91] though all other values of f_6 are allowed.

Trivalent polyhedra must always contain an even number of vertices (n=V), as the number of edges (E=3n/2) must be an integer. Inserting the total number of faces ($F = 12 + f_6$) and edges into Eq. (1), we obtain the following relationship between the number of vertices and the number of hexagonal faces of a fullerene:

$$f_6 = n/2 - 10 \tag{31a}$$
$$n = 20 + 2f_6 \tag{31b}$$

Since f_6 can take any integer value apart from 1, fullerenes C_n are possible with n taking any even number value greater than or equal to 20 – except 22.

Fowler has stated that for $n \geq 28$ each fullerene has at least two distinct isomers and that the number of isomers rises very rapidly with n [12].

7.3.2
Counting Isomers

The elegant geometric, group and graph theoretical methods used in the enumeration of fullerene isomers have been reviewed by Fowler [12] and the isomers of a large number of fullerenes are listed in a compendium of structures compiled by Fowler and Manolopoulos [90].

Fowler and co-workers adapted the Coxeter-Goldberg method of constructing icosahedral deltahedra by triangulating a plane and then wrapping it up into a 3-D polyhedron to generate fullerenes as the duals of the deltahedra thus produced [92, 93]. This led to the generation of two series of icosahedral fullerenes (with $n = 60k$ and $n = 60k + 20$, where $k = 0, 1, 2, ...$) which were shown to be closed shell and open shell respectively (see Sect. 7.3.3).

The most elegant method for enumerating fullerene isomers is the "Spiral Algorithm" of Manolopoulos et al. [94a], in which the fullerene is conceptually peeled as a single 1-D strip of pentagons and hexagons whose order are recorded as a spiral sequence: 56666656..... etc.The method can be inverted to enable isomer counting by constructing all possible sequences of twelve fives and $(n/2 - 10)$ sixes and checking for closure and uniqueness. From the spiral sequence, the connectivity pattern (i.e. the adjacency matrix) can be constructed, which enables the generation of topological co-ordinates and hence the 3-D fullerene structure. It has been proven that all fullerenes with C_5 or C_6 rotation axes have at least one spiral [94b], for example icosahedral C_{60} has three. Although it appears that the vast majority of fullerene isomers have at least one spiral, some counter examples have been found. The smallest fullerene without a spiral code is an isomer of C_{380} [94c].

Finally, Fowler et al. have outlined the fascinating possibility of "bond stretch isomerism" (or π-tautomerism) in the fullerenes [95]. This may occur when, for a given connectivity (σ-framework), more than one Kekulé structure is consistent with the point group symmetry of the cluster and may be a local minimum on the potential energy hypersurface. (The terms 'σ' and 'π' are used here to indicate the symmetry of orbitals with respect to the bonds (edges) of the cluster. Thus, the L_p^σ orbitals give rise to MOs which have local π-symmetry with respect to the cluster edges.) Calculations on the D_2-symmetry isomer of C_{28}, for example, indicate the presence of three enantiomeric pairs of minima, corresponding to different Kekulé structures [95]. Although bond-stretch fullerene isomers have not yet been detected experimentally, the realisation of this elegant and imaginative hypothesis is eagerly awaited.

7.3.3
Open or Closed Shell?

Since each carbon atom has one electron in its radial p-orbital, an n-vertex fullerene can only be closed shell (i.e. possess no unpaired electrons) if n is even –

though this condition is also necessary for the trivalent polyhedron to be geometrically closed. However, n being even is not a sufficient condition for a closed shell fullerene cluster since there remains the possibility of degenerate or pseudo-degenerate frontier orbitals giving rise to open shell configurations for n electrons.

An example of this situation occurs for the smallest possible fullerene, dodecahedral C_{20}, with no hexagonal faces. Experiments indicate that the observed isomers of C_{20} do not include the fullerene and this has been attributed to the cluster having an open shell [48]. On the basis of Hückel (and also more sophisticated) calculations, the electronic configuration of 20-(p^{π})-electron dodecahedral C_{20} has been determined as $(S^{\sigma}: a_{1g})^2\,(P^{\sigma}; t_{1u})^6\,(D^{\sigma}; h_g)^{10}\,(F^{\sigma}; g_u)^2$ [48].

If fullerenes were truly spherical, with a significant degree of electron delocalisation, then in TSH terms there would be no null or redundant functions and closed shells would correspond to the completion of whole L_p^{σ} manifolds. By analogy with atomic orbitals, this would lead to closed shells for neutral n-vertex fullerenes with $N_e = n = 2(t+1)^2$ electrons. This spherical model predicts closed shells at $N_e = 32, 50, 72$ etc. The spherical model is therefore too exclusive as it does not account for the closed shell nature (verified by MO calculations) of the original fullerene, C_{60} itself! The reason for this failure is the fact that the true fullerenes belong to finite point groups, so that the L_p^{σ} manifolds are split. There are also holes in some of the higher manifolds because polyhedra with pentagonal and hexagonal faces do not represent good coverages of the cluster sphere.

The necessary criteria for "properly closed" shell n-vertex fullerenes (n even) are [11, 12]:

(i) the spectrum of L_p^{σ} orbitals has n/2 low lying (bonding) and n/2 high lying (antibonding) components, and
(ii) orbitals n/2 and (n/2)+1 are not degenerate or pseudo-degenerate (i.e. there is a significant HOMO-LUMO separation).

Fowler has also identified "pseudo-closed shell" fullerenes (where there is a HOMO-LUMO gap, but the LUMO is weakly bonding and low-lying) and "meta-closed shell" fullerenes (where there is a HOMO-LUMO gap, but the HOMO is antibonding and relatively high-lying) [11, 12]. These possibilities, along with the open shell isomers, will generally be less stable than properly closed shell isomers.

A number of methods for predicting and generating properly closed shell fullerenes have been described. The most significant work in this area is due to Fowler [11, 12], using group and graph theoretical methods, underpinned by Hückel, semi-empirical, or ab initio MO calculations. For an arbitrary fullerene geometry, however, a properly closed electronic shell is an exception rather than the rule: Fowler has shown that of the first ~1.5 million fullerene isomers only 17 are properly closed [12]!

In 1986, Fowler [92] introduced the "Leapfrog" construction, whereby a family of closed shell cluster geometries is derived from a single precursor structure via a process which involves omnicapping (capping all the faces) of the initial cluster polyhedron (P) to produce an intermediate (deltahedral)

$$P \xrightarrow{\text{omnicap}} O \xrightarrow{\text{dual}} L$$

Scheme 1. The Leapfrog construction

polyhedron O and subsequently generating the "Leapfrog" trivalent polyhedron L as the dual of O (as shown in Scheme 1).

If the parent polyhedron has n vertices, the Leapfrog polyhedron L (so called because its generation proceeds via a non-trivalent intermediate) has 3n vertices [97]. For parent fullerenes, with $n=20+2k$ ($k=0, 2, 3, 4$ ...) vertices, this gives rise to Leapfrog fullerenes with $60+6k$ vertices [96]. The Leapfrog process can be continued to generate families of clusters with 3n, 9n, 27n etc. vertices from a single n-vertex parent. For icosahedral clusters, those with $60k+20$ atoms are open shell, while for their Leapfrogs, with $180k+60$ atoms (including buckminsterfullerene, $k=0$), $540k+180$ atoms etc. are closed shell. In summary, the importance of the Leapfrog construction is that, as well as being a method of generating fullerene structures, it always generates properly closed shell fullerenes C_{3n}, irrespective of the open or closed shell nature of the parent polyhedron P (C_n) [11, 12].

Fowler and Steer showed that the symmetries spanned by the lowest 3n/2 L_p^σ orbitals of the Leapfrog L, are the same as those spanned by the 3n/2 edges of the polyhedron P and that these 3n/2 MOs will therefore have bonding character [96]. This is essentially a TEO-type approach. In group theoretical terms [44–46, 53], the Leapfrog construction may be expressed as [96]

$$\Gamma_{\text{bond}}(L) = \Gamma_{\text{edges}}(P) \tag{32a}$$
$$\Gamma_b^\sigma(L) = \Gamma^\sigma(E) \tag{32b}$$

where E is the edge-figure polyhedron of P. Similarly the 3n/2 antibonding L_p^σ orbitals of L are topologically equivalent to (and have the same symmetries as) half of the $L^\pi/\bar{L}^\pi$ functions of E (those oriented along the edges of the original polyhedron P):

$$\Gamma_{a\text{-}b}^\sigma(L) = {}^1/_2\, \Gamma^\pi(E) \tag{33}$$

Fowler and Ceulemans [97] later showed that the six lowest-lying LUMOs of Leapfrog fullerenes always have symmetries which span the representations of the 3-D cartesian displacements and rotations:

$$\Gamma_{n/2+1..n/2+7} = \Gamma_{x,y,z} \oplus \Gamma_{Rx,Ry,Rz} \tag{34}$$

By recognising the reciprocal nature of capping and truncation, Johnston presented an alternative derivation of the Leapfrog construction [98]. He showed that taking the dual (D) of P and then omnitruncating it generates the same polyhedron (L) as omnicapping P then taking the dual, as shown in Scheme 2:

It is easily shown [98] that omnitruncation of any polyhedron (in this case D) generates a polyhedron with twice as many vertices as D has edges and three times as many edges as D. The closed shell nature of L is again rationalised in terms of the topological equivalence between the bonding half of the L^σ mani-

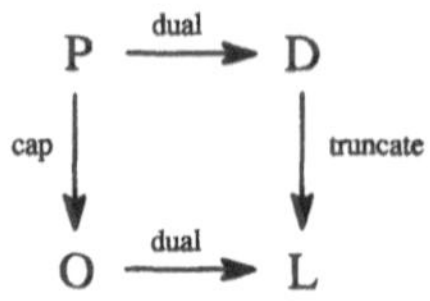

Scheme 2. Alternative Leapfrog pathways

fold of L and the 3n/2 L^σ orbitals of the edge figure E of D (or P, since they share the same edge-figure). Further topological and group theoretical aspects of truncation and their possible significance for cluster bonding were also presented [98].

Fowler and colleagues [99–102] have subsequently shown that all Leapfrog fullerenes are properly closed shells, with strictly bonding HOMOs and anti-bonding LUMOs. Related algorithms, which result in the generation of closed shell polyhedra via vertex doubling and quadrupling, have been presented [93], along with rules for generating closed shell fullerene-like carbon cylinders ("*nano*-tubes") [103, 104].

Although an exhaustive search of all fullerene isomers with $20 \leq n \leq 100$ showed that the Leapfrog and cylinder rules accounts for all isomers with properly closed shells, at larger values of n (e.g. C_{112}), isomers can be found which have properly closed shells but which cannot be generated by a Leapfrog-type algorithm [90].

7.3.4
Isomer Stability

The relative stabilities of fullerene isomers are governed by two factors: the energies of the n/2 occupied L_p^σ MOs and the strain energy of the sp^2 framework. The total L_p^σ energy is related to the open or closed shell nature of the fullerene, since properly closed shell clusters have large HOMO-LUMO gaps (Δ) and, therefore, the sum of the π-electron energies will be lower. Fowler et al. have indicated, however, that while small Δ values, or open shells, are usually associated with low stability, a large Δ does not always imply greater stability [105], due to the greater importance of the steric preferences of the sp^2 framework.

Considering the strain energy of the sp^2 backbone, the main structural feature governing isomer stability is the "Isolated Pentagon Rule" [86], which states that for a given nuclearity (n) the relative energies of fullerene isomers increase with increasing numbers of adjacent pentagons. Indeed, to date, all of the experimentally characterised fullerenes have isolated pentagons. Fowler and Manolopoulos have shown [90] that isolated pentagons only occur in a small percentage of fullerene isomers. Buckminsterfullerene is the smallest fullerene with isolated pentagons and when $n \geq 70$ there is always at least one isomer with isolated pentagons. Given a choice of isomers with the same number of adjacent pentagons, Raghavachari has postulated that the most stable isomers will have the most even distribution of hexagon environments [106].

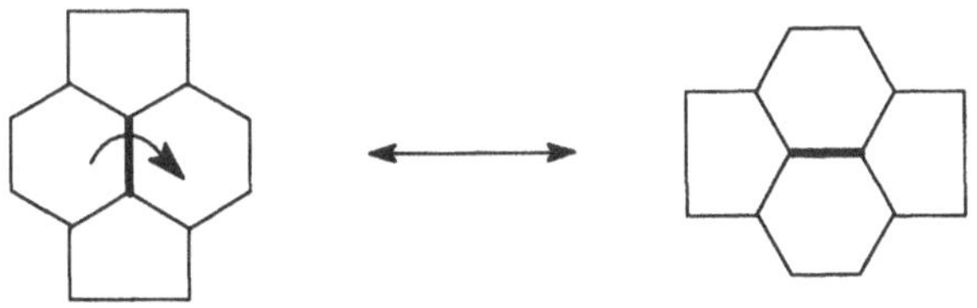

Fig. 9. The Stone-Wales transformation

Fowler and co-workers have recently investigated the extension of these stability rules to include "non-classical" fullerene isomers with square [107] and heptagonal [105] faces, in addition to the usual pentagons and hexagons. For fullerenes with heptagonal faces, Eq. (30b) indicates that there has to be an additional pentagonal face for every heptagon added (since f_5-f_7 = 12). An exciting possibility has been mentioned for C_{62}, for which an isomer with one heptagon is predicted to be more stable than any "classical" fullerene isomer [105].

Finally, there is considerable interest in determining pathways for interconverting fullerene isomers. Fowler and Manolopoulos have mapped out isomerisation pathways for a number of fullerenes [90], based on a mechanism, proposed by Stone and Wales [108], wherein pentagons and hexagons are interconverted via the local rotation of a C_2 fragment, as shown in Fig. 9.

8 Conclusions and Outlook

I hope to have impressed on the reader the fact that, over the past 25 years or so, a wide range of mathematical tools have been brought to bear on various aspects of cluster structure and bonding. Much of this research has been inspired by the synthesis and characterisation of fascinating new cluster types, such as the borane family, the hydrocarbon cages, and, of course, the fullerenes. The fullerene story, highlights par excellence the contributions that mathematical chemists can make to cluster chemistry and the interplay between diverse branches of mathematics, as applied to a family of clusters. As for the future, who knows what novel and exciting clusters may be "discovered" and what mathematical techniques the chemist may have to invent (or borrow) to rationalise their geometric and electronic structures.

Acknowledgment. I am deeply indebted to Professor Mike Mingos for introducing me to the topic of mathematical cluster chemistry and for passing on his enthusiasm and (hopefully) some of his insight. I also wish to thank Professor Bruce King and Dr. David Manolopoulos for helpful suggestions and comments on this review.

9 References

1. Mingos DMP, Johnston RL (1987) Structure & Bonding 68:29
2. King RB (1992) Acc Chem Res 25:247
3. Hoffmann R, Schilling BER, Bau R, Kaesz HD, Mingos DMP (1978) J Amer Chem Soc 100:6088

4. Wade K (1971) J Chem Soc Chem Commun 792
5. Wade K (1976) Adv Inorg Chem Radiochem 18:1
6. Mingos DMP (1972) Nature Phys Sci 236:99
7. Mingos DMP (1974) J Chem Soc Dalton Trans 133
8. Mingos DMP, Wales DJ (1990) Introduction to cluster chemistry. Prentice Hall, Englewood Cliffs, NJ
9. Cundy HM, Rollett AP (1981) Mathematical models, 3rd edn. Tarquin, Norfolk
10. Williams R (1979) The geometrical foundation of natural structure (a source book of design). Dover, New York
11. Fowler PW (1993) In: Kroto HW, Walton DRM (eds.) The fullerenes: new horizons for the chemistry, physics and astrophysics of carbon. Cambridge University Press, Cambridge, p39
12. Fowler PW (1996) Contemp Phys 37:235
13. Wells AF (1977) Three-dimensional nets and polyhedra. Wiley, New York
14. Biggs NL (1974) Algebraic graph theory. Cambridge University Press, London
15. Wilson RJ (1985) Introduction to graph theory, 3rd edn. Longman, New York
16. King RB, Rouvray DH (1977) J Amer Chem Soc 99:7834
17. Balaban AT (ed) (1976) Chemical applications of graph theory. Academic Press, London
18. Hückel E (1931) Z Physik 70:204
19. (a) Lee S-Y, Luo Y-L, Yeh Y-N (1991) J Clust Sci 2:105; (b) Balasubramanian K (1994) Chem Phys Lett 224:325; (c) Balaban AT (1995) J Chem Inf Comp Sci 35:339; (d) Trinajstic N, Randic M, Klein DJ, Babic D, Mihalic Z (1995) Croat Chim Acta 68:241
20. a) Zhao M, Gimarc BM (1993) Inorg Chem 32:4700; (b) Joseph J, Gimarc BM, Zhao M (1993) Polyhedron 12: 2841; (c) Zhao M, Gimarc BM (1995) Polyhedron 14:1315; (d) Gimarc BM, Zhao M (1996) Inorg Chem 35:825
21. Cyrot-Lackman F (1967) J Phys Chem Solids 29:1235
22. (a) Burdett JK (1987) Structure & Bonding 65:29; (b) Burdett JK (1988) Acc Chem Res 21:189
23. Rousseau R, Lee S (1994) J Chem Phys 101:10753
24. King RB, Dai B, Gimarc BM (1990) Inorg Chim Acta 167:213
25. Knight WD, de Heer WA, Clemenger K, Saunders WA (1985) Solid State Commun 53:445
26. Knight WD, Clemenger K, de Heer WA, Saunders WA, Chou MY, Cohen ML (1984) Phys Rev Lett 52: 2141
27. Bjørnholm S (1994) Shell structure in atoms, nuclei and in metal clusters. In: Haberland H (ed) Clusters of atoms and molecules I. Springer series in chemical physics 52. Springer, Berlin Heidelberg New York, p141
28. Chapman D, Waddington TC (1962) Trans Faraday Soc 58:1679
29. Hoffmann R, Gouterman M (1962) J Chem Phys 36:2189
30. Waddington TC (1967) Trans Faraday Soc 63:1313
31. Hoffman DK, Ruedenberg K, Verkade JG (1977) Structure & Bonding 33:57
32. Stone AJ (1980) Mol Phys 41:1339
33. Stone AJ (1981) Inorg Chem 20:563
34. Stone AJ, Alderton MJ (1982) Inorg Chem 21:2297
35. Stone AJ (1984) Polyhedron 3:2051
36. Johnston RL, Mingos DMP (1986) Inorg Chem 25:1661
37. Brint P, Cronin JP, Seward E (1983) J Chem Soc Dalton Trans 675
38. Johnston RL, Mingos DMP (1989) Theor Chim Acta 75:11
39. Fowler PW, Porterfield WW (1985) Inorg Chem 24:3511
40. Wales DJ, Stone AJ (1987) Inorg Chem 25:3845
41. (a) Wales DJ, Stone AJ (1987) Inorg Chem 28:2754; (b)Mingos DMP, Johnston RL (1988) Polyhedron 7:2437
42. Gimarc BM, Ott JJ (1986) Inorg Chem 25:83, 2708
43. King RB (1993) J Math Chem 12:9
44. Redmond DB, Quinn CM, McKiernan JG (1983) J Chem Soc Faraday Trans II 79:1791
45. Quinn CM, McKiernan JG, Redmond DB (1983) Inorg Chem 22:2310

46. Quinn CM, McKiernan JG, Redmond DB (1984) J Chem Ed 61:569, 572
47. Fowler PW (1985) Polyhedron 4:2051
48. Fowler PW, Woolrich J (1986) Chem Phys Lett 127:84
49. Johnston RL, Mingos DMP (1987) J Chem Soc Dalton Trans 647
50. Johnston RL, Mingos DMP (1987) J Chem Soc Dalton Trans 1445
51. Ceulemans A (1985) Mol Phys 54:161
52. Cotton FA (1971) Chemical applications of group theory, 2nd edn. Wiley, New York
53. Fowler PW, Quinn CM (1986) Theor Chim Acta 70:333
54. Heitler H, London F (1927) Z Phys 44:455
55. Longuet-Higgins HC (1949) J Chim Phys 46:275
56. Duffey GH (1953) J Chem Phys 21:761
57. Eberhardt WH, Crawford B, Lipscomb WN (1954) J Chem Phys 22:989
58. Lipscomb WN (1963) Boron hydrides. Benjamin, New York
59. Wade K (1971) Electron deficient compounds. Nelson, London
60. Lennard-Jones JE (1949) Proc Roy Soc (London) A 198:1, 14
61. (a) Kettle SFA, Tomlinson V (1969) J Chem Soc A: 2002, 2007; (b) Kettle SFA, Tomlinson V (1969) Theor Chim Acta 14:175
62. (a) Longuet-Higgins HC, Roberts M. de V. (1954) Proc Roy Soc (London) A224:336; (b) Longuet-Higgins HC, Roberts M. de V. (1955) Proc Roy Soc (London) A230:110
63. Sheldon JC (1960) J Chem Soc 1007, 3106
64. Crossmann LD, Olsen DP, Duffey GH (1963) J Chem Phys 38:73
65. Kettle SFA (1965) Theor Chim Acta 3:211
66. Cotton FA, Haas TE (1964) Inorg Chem 3:10
67. Kettle SFA (1966) Theor Chim Acta 4:150
68. Kettle SFA (1964) J Chem Soc A:314
69. Lauher JW (1978) J Amer Chem Soc 100:5305
70. Johnston RL, Lin Z, Mingos DMP (1989) New J Chem 13:33
71. Bicerano G, Marynick DS, Lipscomb WN (1978) Inorg Chem 17:2041, 3443
72. Johnston RL, Mingos DMP (1986) Polyhedron 5:2059
73. Johnston RL, Mingos DMP (1986) Inorg Chem 25:3321
74. Johnston RL, Mingos DMP, Sherwood P (1991) New J Chem 15:831
75. Ternansky RJ, Balogh DW, Paquette LA (1982) J Amer Chem Soc 104:4503
76. Schulman JM, Fischer CR, Solomon CR, Venanzi TJ (1978) J Amer Chem Soc 100:2949
77. Galasso V (1994) Chem Phys 184:107
78. Heilbronner E, Jones TB, Krebs A, Malsch D-D, Maier G, Pocklington J, Schmelzer A (1980) J Amer Chem Soc 102:564
79. Johnston RL, Mingos DMP (1985) J Organomet Chem 280:407
80. Johnston RL (1986) DPhil Thesis, University of Oxford
81. Schulman JM, Disch RL, Miller MA, Peck RC (1987) Chem Phys Lett 141:45
82. Stanton JF, Gauss J, Watts JD, Bartlett RJ (1991) J Chem Phys 94:4334
83. Johnston RL, Mingos DMP (1985) J Organomet Chem 280:419
84. Mingos DMP, Lin Z (1989) Structure & Bonding 71:1
85. Kroto HW, Heath JR, O'Brien SC, Curl RF, Smalley RE (1985) Nature 318:162
86. Kroto HW (1987) Nature 329:529
87. Krätschmer W, Lamb LD, Fostiropolous K, Huffman DR (1990) Nature 347:354
88. Kroto HW, Allaf AW, Balm SP (1991) Chem Rev 91:1213
89. Kroto HW, Walton DRM (1993) (eds.) The fullerenes: new horizons for the chemistry, physics and astrophysics of carbon. Cambridge University Press, Cambridge
90. Fowler PW, Manolopoulos DE (1995) An atlas of fullerenes. Oxford University Press, Oxford
91. Grünbaum B (1967) Convex polytopes. Wiley-Interscience, New York
92. Fowler PW (1986) Chem Phys Lett 131:444
93. Fowler PW, Cremona JE, Steer JI (1988) Theor Chim Acta 73:1
94. (a) Manolopoulos DE, May JC, Down SE (1991) Chem Phys Lett 181:105; (b) Fowler PW, Manolopoulos DE, Redmond DB, Ryan RP (1993) Chem Phys Lett 202:371; (c) Manolopoulos DE, Fowler PW (1993) Chem Phys Lett 204:1

95. Austin SJ, Baker J, Fowler PW, Manolopoulos DE (1994) J Chem Soc Perkin Trans 2 2319
96. Fowler PW, Steer JI (1987) J Chem Soc Chem Commun 1403
97. Fowler PW, Ceulemans A (1995) J Phys Chem 99:508
98. Johnston RL (1991) J Chem Soc Faraday Trans 87:3353
99. Manolopoulos DE, Woodall DR, Fowler PW (1992) J Chem Soc Faraday Trans 88:2427
100. Fowler PW, Redmond DB (1992) Theor Chim Acta 83:367
101. Fowler PW, Pisanski T (1994) J Chem Soc Faraday Trans 90:2865
102. Manolopoulos DE, Fowler PW (1992) J Chem Phys 96:7603
103. Fowler PW (1990) J Chem Soc Faraday Trans 86:2073
104. Fowler PW (1992) J Chem Soc Faraday Trans 88:2631
105. Fowler PW, Heine T, Mitchell D, Orlandi G, Schmidt R, Seifert G, Zerbetto F (1996) J Chem Soc Faraday Trans 92:2203
106. Raghavachari K (1992) Chem Phys Lett 190:397
107. Fowler PW, Heine T, Manolopoulos DE, Mitchell D, Orlandi G, Schmidt R, Seifert G, Zerbetto F (1996) J Phys Chem 100:6984
108. Stone AJ, Wales DJ (1986) Chem Phys Lett 128:501

Metal-Metal Interactions in Transition Metal Clusters with π-Donor Ligands

Zhenyang Lin and Man-Fai Fan

Department of Chemistry, Hong Kong University of Science and Technology, Clear Water Bay, Kowloon, Hong Kong. *E-mail address: chzlin@usthk.ust.hk*

A variety of structural types of transition metal clusters containing mainly π-donor ligands are reviewed. Their structure and bonding are analyzed using a simple and unified molecular orbital approach based on the consideration of metal-metal bonding as arising from the interaction of the frontier orbitals of individual metal fragments derived from their local ML_n coordination. This kind of "local metal frontier orbital" approach allows us to construct metal-metal orbital interaction diagrams in a convenient way. Using these molecular orbital diagrams, relevant optimal electron counts for major structural types are discussed. It is emphasized that the "t_{2g}" set fragment orbitals, which are nonbonding in carbonyl clusters, play the crucial role in metal-metal bonding for clusters with π-donor ligands. Bonding characteristics for Mo(W)-Fe-S, Ni-S(Se) and related clusters are also discussed.

Keywords: Metal-metal bonding; halide and sulfide clusters

1 Introduction

During the last several decades, molecular cluster compounds have attracted a great deal of attention because of their relevance to catalysts, to structural and functional models of metallaenzymes, and to solid state properties such as superconductivity [1–8]. In 1971, Williams [9] and Wade [10] established the relationships between structure and the closed shell electronic requirements for boranes and carboranes, i.e., the famous n + 1 rule. One year later, both Wade [12] and Mingos [11] recognized the important bonding similarities between these main group clusters and a large number of transition metal carbonyl clusters. Since then, the geometries of these clusters have been systematized by a set of rules described collectively as the Polyhedral Skeletal Electron Pair Theory (PSEPT) which accounts for the relationship between the skeletal structure and the total number of valence electrons [13]. The theoretical basis for this approach has been underpinned by the Tensor Surface Harmonic (TSH) theory [14] and the isolobal analogy [15]. While these rules have proved to be extremely successful in clarifying the structure and bonding of transition metal carbonyl clusters, they are not so applicable to metal clusters containing mainly π-donor ligands. For example, many face- and edge-bridged octahedral clusters [16] with the general formula $M_6(\mu_3\text{-}X)_8L_6$ and $M_6(\mu_2\text{-}X)_{12}L_6$ (M = transition metal; X = bridging π-donor ligands; L = terminal ligand) have electron counts of 84 (24 metal-metal bonding d electrons + 60 ligand electrons) and 76 (16 metal-metal bonding d electrons + 60 ligand electrons) respectively, while an octahedral transition metal carbonyl cluster normally conforms to the n+1 rule and contains 86 electrons (14 metal-metal skeletal bonding electrons), e.g., $Os_6(CO)_{18}^{2-}$ [17].

Parallel to the development of transition metal carbonyl clusters, a large number of non-carbonyl clusters have been synthesized and structurally characterized. These clusters, in general, contain earlier transition metals and are associated with (bridging) ligands such as oxo, sulfido, halide and alkoxyl groups. The well-known triangular cluster $Re_3Cl_{12}^{3-}$, which was discovered by Cotton et al. [18] and led to the recognition of metal-metal (multiple) bonds, is one of this type. The discussion of these non-carbonyl clusters has been focused on their electronic structures, which are closely related to the study of metal-metal bonds. In this article, we will attempt to provide a systematic approach in understanding the structure and bonding for those clusters containing mainly π-donor ligands. A simple and unified molecular orbital picture will be used to discuss the relationship between structure and electron count, based on the consideration of metal-metal bonding as arising from the interaction of the frontier orbitals of-individual metal fragments derived from their local ML_n coordination. This kind of "local metal frontier orbital" approach has been used by various authors for metal-metal interactions in specific types of clusters [19]. The main purpose here is to demonstrate the versatility of this approach in rationalizing the electronic structures of a wide variety of clusters with π-donor ligands.

In the next section, the frontier orbitals of a metal carbonyl and halide fragments will be compared, laying down the basis of orbital interactions for the whole discussion. This is followed by considering various types of clusters, with emphasis on the electron counting rules derived from the qualitative MO diagrams. There is also a section for the biologically important Mo-Fe-S clusters, which are deemed to have weak metal-metal interactions Although we do not intend to provide a comprehensive survey of transition metal clusters with π-donor ligands, selected examples will be provided for each structural type for illustration. Carbonyl clusters will be used only when comparisons with analogous halide clusters are necessary. To simplify the discussions, only clusters with symmetrically disposed ligands will be considered, and acetato ($CH_3CO_2^-$) and cyclopentadienyl ($C_5H_5^-$) ligands are considered as the equivalent of two and three terminal ligands respectively. The focus of the present article will be on metal clusters with local ML_n ($n = 4, 5, 6,$ and 7) environments, for which the separation of metal-ligand σ and π bonding is easily understood. It should be noted that the term "carbonyl or halide clusters" refers to clusters with generic π-acceptor or π-donor ligands.

2
Bonding Characteristics of Carbonyl and Halide Clusters

Main group boranes or carboranes and transition metal carbonyl *closo* deltahedral clusters are characterized by $4N+2$ and $14N+2$ valence electrons, respectively (N indicates the number of skeletal atoms in the clusters). This bonding similarity has been explained in terms of the isolobal relationship between AH (A: boron or carbon) and ML_3 (M: transition metal atom; L: carbonyl or other π-acid ligand) fragments. Both fragments have three frontier fragment orbitals, shown in Fig. 1, for skeletal bonding in clusters. Excluding the three frontier orbitals, the occupation of remaining bonding and non-bonding orbi-

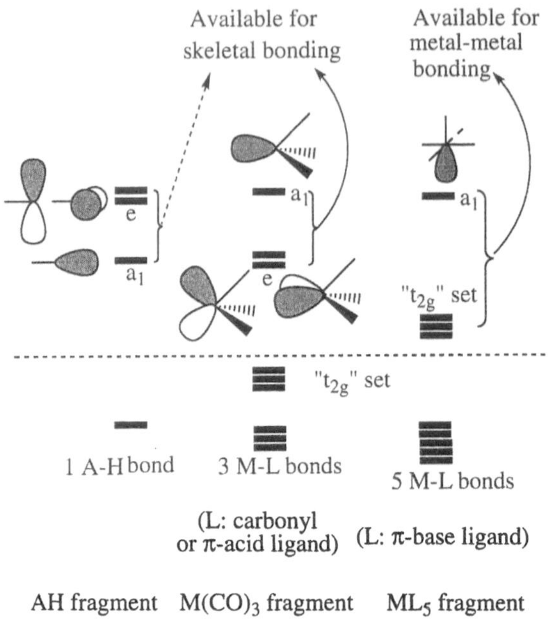

Fig. 1. Frontier fragment orbitals for AH, $M(CO)_3$ and ML_5 fragments

tals leads to 2N electrons for main group fragments and 12N electrons for transition metal fragments. The difference between main group and transition metal cluster electron counts is, therefore, 10N when both clusters have the same skeletal structure.

In transition metal carbonyl clusters, the so-called "t_{2g}" set (see Fig. 1), which are mainly derived from metal's d orbitals, are essentially non-bonding, i.e., almost inert toward participation of metal-metal bonding. The inertness of these "t_{2g}" sets results from their significant stabilization by the carbonyl π^* orbitals through the well-known Dewar-Chatt back-bonding. In transition metal clusters containing π-donor ligands, however, the "t_{2g}" fragment orbitals are crucial in metal-metal bonding. They become active in the formation of metal-metal bonding because they are significantly destabilized by π-donating orbitals from ligands. Figure 1 also illustrates the frontier orbitals of a square pyramidal (C_{4v}) ML_5 fragment, which have been shown to play important roles in the metal-metal bonding for a large number of face- and edge-bridged octahedral halide and chalcogenide clusters [16].

3
Dinuclear Metal Clusters

Since the discovery of the quadruple bond in $Re_2Cl_8^{2-}$, dinuclear metal clusters have served as the prototype for the study of metal-metal interactions, with the pioneering work being done by Cotton and Walton [5]. In contrast to dinuclear metal carbonyl complexes (e.g., $Mn_2(CO)_{10}$) where the presence of even an M=M double bond is rare, bond orders of 2–4 are commonplace for dinuclear halide clusters, and an M-M single bond is only observed when bridging ligands are also involved. It is now known that the destabilization of the "t_{2g}" set by π-donor ligands is necessary to support M-M bond orders higher than two. In addition to these "unsupported" dinuclear clusters, bridging π-donor ligands result in the formation of edge- and face-sharing bioctahedral complexes with interesting structural, bonding and reactivity characteristics.

As numerous reviews [5, 19, 20] on these types of dinuclear complexes have been published, notably by Cotton and his coworkers, here we only provide a brief summary of the metal-metal interactions and electron counting, using the local frontier orbital approach to unite the bonding descriptions of all the structural types. Monobridged dinuclear complexes are excluded from the discussion since the metal-metal interaction is not well defined.

3.1
Dinuclear Clusters Based on ML_4 or ML_5 Fragments

Structures of this type are shown in Fig. 2. Each transition metal center has locally five or four metal-ligand σ bonds. For an ML_5 fragment, the frontier orbitals are a σ-type hybrid and the "t_{2g}" orbitals (see Fig. 1). For a square planar ML_4 fragment, the corresponding frontier orbitals are now a d_z2 orbital, instead of the σ-type hybrid, and also the "t_{2g}" orbitals (see also Fig. 2). As shown above,

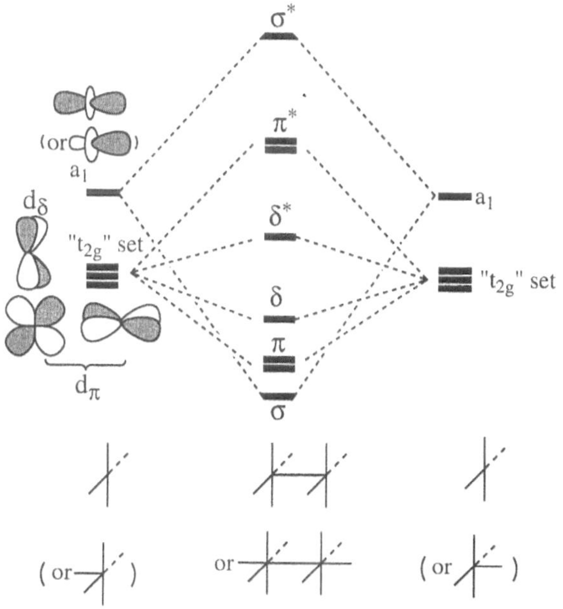

Fig. 2. Orbital interaction diagram for dinuclear clusters based on ML_4 or ML_5 fragments

the "t_{2g}" orbitals are very important in metal-metal bonding for clusters containing π-donating ligands. The metal-metal orbital interaction between the two fragments gives rise to four bonding ($1\sigma+2\pi+1\delta$) and four antibonding ($1\sigma^*+2\pi^*+1\delta^*$) molecular orbitals, and is illustrated in Fig. 2. The occupation of the four M-M bonding molecular orbitals gives the optimal metal total d-electron count of eight, forming the famous quadruple bonds. Deviation from this optimal occupation number results in a smaller bond order, and hence longer M-M bond lengths and weaker bond strengths. Some typical examples with this structural type are given in Table 1 together with their d electron counts. The correlation between M-M electron count and bond length is immediately evident.

The occupation of δ bonding orbital in the classic complex $Re_2Cl_8^{2-}$ results in an eclipsed conformation. Occupation of the δ^* orbital cancels the effect of δ bonding, and such complexes are expected to have a staggered conformation due to the steric effect of the ligands. However, the energy splitting of the δ and δ^* orbitals is small such that, for 10-electron complexes, in which the δ^* orbital is occupied, both the eclipsed ($Re_2Cl_4(PMe_3)_4$ [28]) and staggered ($Os_2Cl_8^{2-}$ [31]) conformations are observed.

For late transition metals, the metal d electron counts always exceed the optimal electronic configuration of eight. The d orbitals, which are crucial in forming metal-metal bonds, become contracted toward the right end of the transition series. Hence the overlap of d orbitals becomes poor, and the corresponding bonding and antibonding orbitals do not split as much as in the earlier transition metals, thus allowing the δ^* and π^* orbitals to be occupied. Group 10 clusters with 16 and 18 metal d electrons do not have net metal-metal bonding, and their stabilities are the results of chelating effects from the bridging bidentate ligands. For dinuclear clusters with 18 metal d electrons, one metal-ligand antibonding orbital is also occupied.

Dinuclear clusters with local ML_4 or ML_5 coordination have a minimum electron count of eight. However, an important type of Group 6 alkoxide clusters have a total of six d electrons (see Sect. 3.5).

3.2 Dinuclear Clusters Based on Edge-Sharing Bioctahedra

An edge-sharing bioctahedron (ESBO) has a local octahedral ML_6 coordination (Fig. 3). In other words, each metal center has the "t_{2g}" fragment orbitals available for the metal-metal interaction. The relevant orbital interaction diagram is shown in Fig. 3. Compared with unbridged dinuclear clusters, an ESBO only has one π orbital available, rather than two in unbridged clusters. In the orbital level diagram (Fig. 3), the ordering of δ and δ^* is not clearly defined. The δ orbital can be destabilized through its interaction with the bridging ligand's π orbitals (see **1**) while the δ^* cannot find a symmetry-adapted combination from the bridging ligands. As a result, the δ^* orbital becomes lower in energy than the δ orbital for the majority of ESBOs with π-donor ligands, and the maximum bond order is thus two instead of three, achieved by electronic configurations of $\sigma^2\pi^2$ and $\sigma^2\pi^2\delta^{*2}\delta^2$. Some typical examples are listed in Table 1 together with their

Table 1. Representative examples of dinuclear transition metal clusters

Cluster Compound	Number of metal-metal bonding electrons	M-M bond distance (Å)	Electronic configuration	Reference
Dinuclear Clusters Bases on ML_4 or ML_5 Fragments				
$Tc_2Cl_8^{2-}$	8	2.151	$\sigma^2\pi^4\delta^2$	[21]
$[Tc_2(\mu_2\text{-acetato})_4]^{1+}$	9	2.117	$\sigma^2\pi^4\delta^2\delta^{*1}$	[22]
$[Tc_2(\mu_2\text{-acetato})_4Cl_2]^{1-}$	9	2.126	$\sigma^2\pi^4\delta^2\delta^{*1}$	[23]
$Tc_2Cl_4PMe_2Ph)_4$	10	2.128	$\sigma^2\pi^4\delta^2\delta^{*2}$	[24]
$Re_2(\mu_2\text{-acetato})_4Cl_2$	8	2.224	$\sigma^2\pi^4\delta^2$	[25]
$Re_2Br_8^{2-}$	8	2.226	$\sigma^2\pi^4\delta^2$	[26]
$Re_2(NCS)_8^{2-}$	8	2.270	$\sigma^2\pi^4\delta^2$	[27]
$Re_2Cl_4(PMe_3)_4$	10	2.247	$\sigma^2\pi^4\delta^2\delta^{*2}$	[28]
$[Ru_2(\mu_2\text{-acetato})_4(H_2O)_2]^{1+}$	11	2.248	$\sigma^2\pi^4\delta^2\delta^{*2}\pi^{*1}$	[29]
$Ru_2(\mu_2\text{-acetato})_4(thf)_2$	12	2.261	$\sigma^2\pi^4\delta^2\delta^{*2}\pi^{*2}$	[30]
$Os_2Cl_8^{2-}$	10	2.182	$\sigma^2\pi^4\delta^2\delta^{*2}$	[31]
$Co_2(NCS)_8^{4-}$	14	2.736	$\sigma^2\pi^4\delta^2\delta^{*2}\pi^{*4}$	[32]
$Rh_2(\mu_2\text{-acetato})_4(Sthf)_2$	14	2.413	$\sigma^2\pi^4\delta^2\delta^{*2}\pi^{*4}$	[33]
$Ni_2(\mu_2\text{-dithioacetato})_4$	16	2.564	$\sigma^2\pi^4\delta^2\delta^{*2}\pi^{*4}\sigma^{*2}$	[34]
$Pd_2(\mu_2\text{-dmpe})_4Br_2$	18	2.603	a	[35]
$[Pd_2(\mu_2\text{-acetato})_4(H_2O)_2]^{2+}$	14	2.578	$\sigma^2\pi^4\delta^2\delta^{*2}\pi^{*4}$	[36]
$Pd_2(\mu_2\text{-dithioacetato})_4$	16	2.767	$\sigma^2\pi^4\delta^2\delta^{*2}\pi^{*4}\sigma^{*2}$	[37]
$Pd_2(\mu_2\text{-dppm})_4Cl_2$	18	2.651	a	[38]
Dinuclear Clusters Based on Edge-Sharing Bioctahedra				
$Ti_2(\mu_2\text{-Cl})_2(thf)_4Cl_4$	2	3.711	σ^2	[39]
$Zr_2(\mu_2\text{-Cl})_2Cl_4(P^nBu_3)_4$	2	3.183	σ^2	[40]
$Zr_2(\mu_2\text{-I})_2I_4(PMe_3)_4$	2	3.388	σ^2	[41]
$Hf_2(\mu_2\text{-Cl})_2Cl_4(PMe_2Ph)_4$	2	3.395	σ^2	[41]
$V_2(\mu_2\text{-S})_2Cl_2(\eta^5\text{-Cp})_2$	2	2.829	σ^2	[42]
$V_2(\mu_2\text{-Cl})_2Cl_4(PMe_3)_4$	4	3.696	$\sigma^2\pi^2$	[43]
$V_2(\mu_2\text{-Cl})_2(PEt_3)_2(\eta^5\text{-Cp})_2$	6	3.245	$\sigma^2\pi^2\delta^{*2}$	[44]
$Nb_2(\mu_2\text{-O})_2(O\text{-}C_6H_3{}^iPr_2)_4$	0	3.050	σ^0	[45]
$Nb_2(\mu_2\text{-S})_2Cl_4(Sthf)_4$	2	2.867	σ^2	[46]
$Nb_2(\mu_2\text{-Cl})_2Cl_4(dppm)_2$	4	2.696	$\sigma^2\pi^2$	[47]
$Ta_2(\mu_2\text{-OEt})_2Cl_4(OEt)_4$	0	3.413	σ^0	[48]
$Ta_2(\mu_2\text{-S})_2Cl_4(PMe_2Ph)_4$	2	2.875	σ^2	[49]
$Ta_2(\mu_2\text{-Cl})_2Cl_4(PMe_2Ph)_4$	4	2.710	$\sigma^2\pi^2$	[50]
$Cr_2(\mu_2\text{-Cl})_2Cl_4(PMe_3)_4$	6	3.590	$\sigma^2\pi^2\delta^{*2}$	[51]
$Mo_2(\mu_2\text{-S})_2S_2(\eta^5\text{-Cp})_2$	2	2.905	σ^2	[52]
$Mo_2(\mu_2\text{-O}^iPr)_2(O^iPr)_4Cl_4$	2	2.731	σ^2	[53]
$[Mo_2(\mu_2\text{-Cl})_2Cl_8]^{2-}$	4	3.808	$\sigma^2\pi^2$	[54]
$[Mo_2(\mu_2\text{-SEt})_2Cl_4](dmpe)_2$	6	2.712	$\sigma^2\pi^2\delta^2$	[55]
$W_2(\mu_2\text{-S})_2S_2(\eta^5\text{-Cp})_2$	2	2.900	σ^2	[56]
$W_2(\mu_2\text{-OMe})_2Cl_4(OMe)_2(MeOH)_2$	4	2.481	$\sigma^2\pi^2$	[57]
$W_2(\mu_2\text{-S})_2(S^tBu)_4(PMe_2Ph)_4$	4	2.736	$\sigma^2\pi^2$	[58]
$W_2(\mu_2\text{-Cl})_2Cl_4(PEt_3)_4$	6	2.740	$\sigma^2\pi^2\delta^{*2}$	[59]
$[Mn_2(\mu_2\text{-Cl})_2Cl_4(H_2O)_4]^{2-}$	10	3.827	$\sigma^2\pi^2\delta^{*2}\delta^2\pi^{*2}$	[60]
$Re_2(\mu_2\text{-OMe})_2(OMe)_8$	4	2.532	$\sigma^2\pi^2$	[61]
$Re_2(\mu_2\text{-Cl})_2Cl_4(dppm)_2$	8	2.616	$\sigma^2\pi^2\delta^{*2}\delta^2$	[62]

Table 1 (continued)

Cluster Compound	Number of metal-metal bonding electrons	M-M bond distance (Å)	Electronic configuration	Reference
$Re_2(\mu_2\text{-}Cl)_2Cl_2(\eta^5\text{-}C_5MeEt_4)_2$	8	2.502	$\sigma^2\pi^2\delta^{*2}\delta^2$	[63]
$[Fe_2(\mu_2\text{-}F)_2F_6(H_2O)_2]^{2-}$	10	3.171	$\sigma^2\pi^2\delta^{*2}\delta^2\pi^{*2}$	[64]
$Ru_2(\mu_2\text{-}Cl)_2Cl_4(P^nBu_3)_4$	10	3.733	$\sigma^2\pi^2\delta^{*2}\delta^2\pi^{*2}$	[65]
$Ru_2(\mu_2\text{-}Cl)_2Cl_2(\eta^6\text{-arene})_2$	12	3.743	$\sigma^2\pi^2\delta^{*2}\delta^2\pi^{*2}\sigma^{*2}$	[66]
$[Os_2(\mu_2\text{-}Cl)_2Cl_8]^{2-}$	8	3.626	$\sigma^2\pi^2\delta^{*2}\delta^2$	[67]
$Os_2(\mu_2\text{-}Br)_2Br_2(\eta^5\text{-}Cp)_2$	10	2.970	$\sigma^2\pi^2\delta^{*2}\delta^2\pi^{*2}$	[68]
$Os_2(\mu_2\text{-}Cl)_2Cl_2(\eta^6Me^iPr\text{-arene})_2$	12	3.741	$\sigma^2\pi^2\delta^{*2}\delta^2\pi^{*2}\sigma^{*2}$	[69]
$Rh_2(\mu_2\text{-}S)_2Cl_2H_2(PPh_3)_4$	10	3.637	$\sigma^2\pi^2\delta^{*2}\delta^2\pi^{*2}$	[70]
$Rh_2(\mu_2\text{-}Cl)_2Cl_4(P^nBu_3)_4$	12	3.744	$\sigma^2\pi^2\delta^{*2}\delta^2\pi^{*2}\sigma^{*2}$	[71]
$Ir_2(\mu_2\text{-}Cl)_2Cl_2(\eta^5\text{-}C_5Me_5)_2$	12	3.770	$\sigma^2\pi^2\delta^{*2}\delta^2\pi^{*2}\sigma^{*2}$	[72]
$[Ni_2(\mu_2\text{-}Cl)_2Cl_6(H_2O)_2]^{4-}$	16	3.604	a	[73]
Dinuclear Clusters Based on Face-Sharing Bioctahedra				
$[Ti_2(\mu_2\text{-}Cl)_3Cl_6]^-$	0	3.394	$(a_1)^0$	[74]
$[Zr_2(\mu_2\text{-}I)_3I_4(PEt_3)_2$	1	3.170	$(a_1)^1$	[75]
$[V_2(\mu_2\text{-}Cl)_3Cl_6]^-$	2	3.325	$(a_1)^2$	[76]
$[V_2(\mu_2\text{-}Cl)_3Cl_4(thf)_2]^-$	4	3.175	$(a_1)^2(e)^2$	[77]
$[Nb_2(\mu_2\text{-}Cl)_3Cl_4(PEt_3)_2]^-$	4	2.707	$(a_1)^2(e)^2$	[78]
$Ta_2(\mu_2\text{-}Cl)(\mu_2\text{-}SEt)_2Cl_4(SEt_2)_2$	3	2.758	$(a_1)^2(e)^1$	[79]
$[Ta_2(\mu_2\text{-}Cl)_2(\mu_2\text{-}SMe_2)Cl_4(SMe_2)_2$	4	2.692	$(a_1)^2(e)^2$	[80]
$[Cr_2(\mu_2\text{-}F)_3F_6]^{3-}$	6	2.771	$(a_1)^2(e)^4$	[81]
$[Mo_2(\mu_2\text{-}I)_3I_4(PMe_3)_2]^-$	6	3.022	$(a_1)^2(e)^4$	[82]
$W_2(\mu_2\text{-}S)(\mu_2\text{-}SEt)_2Cl_4(SMe_2)_2$	4	2.525	$(a_1)^2(e)^2$	[83]
$[W_2(\mu_2\text{-}SPh)_2(\mu_2\text{-}Cl)CL_6]^{2-}$	5	2.519	$(a_1)^2(e)^3$	[84]
$W_2(\mu_2\text{-}SEt)_3Cl_4(SMe_2)_2$	5	2.504	$(a_1)^2(e)^3$	[85]
$[Fe_2(\mu_2\text{-}Cl)_3Cl_6]^{3-}$	12	2.899	$(a_1)^2(e)^4(a_1{}^*)^2(e^*)^4$	[86]
$[Ru_2(\mu_2\text{-}Cl)_3Cl_2(PMe_2Ph)_4$	11	2.994	$(a_1)^2(e)^4(a_1{}^*)^2(e^*)^3$	[87]
$[Ru_2(\mu_2\text{-}Cl)_3(P^nBu_3)_6]^+$	12	3.395	$(a_1)^2(e)^4(a_1{}^*)^2(e^*)^4$	[87]
$[Rh_2(\mu_2\text{-}Cl)_2Cl_4(PEt_3)_2]^-$	12	3.175	$(a_1)^2(e)^4(a_1{}^*)^2(e^*)^4$	[88]
$[Rh_2(\mu_2\text{-}OH)_3Cl_2(PPh_2OH)]^-$	14	3.265	a	[89]
$[Ir_2(\mu_2\text{-}OH)_3(\eta^5\text{-}CpMe_5)_2]^{1+}$	12	3.071	$(a_1)^2(e)^4(a_1{}^*)^2(e^*)^4$	[90]
$[Ni_2(\mu_2\text{-}Cl)_3(thf)_6]^{1+}$	16	2.993	a	[91]
Dinuclear Clusters with Quadraruple Bridging Ligands				
$L_3M(\mu_2\text{-}X)_4ML_3$				
$Nb_2(\mu_2\text{-}Br)_4(\eta^5\text{-}CpMe_5)_2$	4	2.761	$\sigma^2\delta^{*2}$	[92]
$Ta_2(\mu_2\text{-}Br)_4(\eta^5\text{-}CpMe_5)_2$	4	2.768	$\sigma^2\delta^{*2}$	[93]
$Mo_2(\mu_2\text{-}S)_2(\mu_2\text{-}SMe)_2(\eta^5\text{-}CpMe_5)_2$	4	2.573	$\sigma^2\delta^{*2}$	[94]
$Cr_2(\mu_2\text{-}S)_2(\mu_2\text{:}\eta^2\text{-}S_2)(\eta^5\text{-}CpMe_5)_2$	4	2.463	$\sigma^2\delta^{*2}$	[95]
$Mo_2(\mu_2\text{-}S)_2(\mu_2\text{:}\eta^2\text{-}S_2)(\eta^5\text{-}CpMe_5)_2$	4	2.599	$\sigma^2\delta^{*2}$	[96]
$Mo_2(\mu_2\text{-}Cl)_4(\eta^5\text{-}Cp^iPr)_2$	6	2.607	$\sigma^2\delta^{*2}\delta^2$	[97]
$Mo_2(\mu_2\text{:}\eta^2\text{-}SCH{=}CHS)_2(\eta^5\text{-}Cp)_2$	6	2.580	$\sigma^2\delta^{*2}\delta^2$	[98]
$W_2(\mu_2\text{-}Cl)_4(\eta^5\text{-}CpMe_5)_2$	6	2.626	$\sigma^2\delta^{*2}\delta^2$	[99]
$L_4M(\mu_2\text{-}X)_4ML_4$				
$Nb_2(\mu_2\text{-}Cl)_4Cl_4(PMe_2Ph)_4$	2	2.838	σ^2	[100]

Table 1 (continued)

Cluster Compound	Number of metal-metal bonding electrons	M-M bond distance (Å)	Electronic configuration	Reference
$Ta_2(\mu_2\text{-}Cl)_4Cl_4(PMe_3)_4$	2	2.830	σ^2	[101]
$[Mo_2(\mu_2{:}\eta^2\text{-}S_2)_2Cl_8]^{2-}$	2	2.858	σ^2	[102]
Unbridged L_3MML_3 Dinuclear Clusters				
$Mo_2(CH_2Ph)_2(NMe_2)_4$	6	2.200	$(a_1)^2(e)^4$	[103]
$Mo_2(O^iPr)_2(SC_6H_2Me_3)_4$	6	2.230	$(a_1)^2(e)^4$	[104]
$Mo_2(OC(CF_3)_2CH_3)_6$	6	2.230	$(a_1)^2(e)^4$	[105]
$W_2Br_2(NEt_2)_4$	6	2.303	$(a_1)^2(e)^4$	[106]

[a] In these clusters, metal-metal bonding and antibonding molecular orbitals are all accupied. In addition, one metal-ligand antibonding molecular orbital is also occupied.

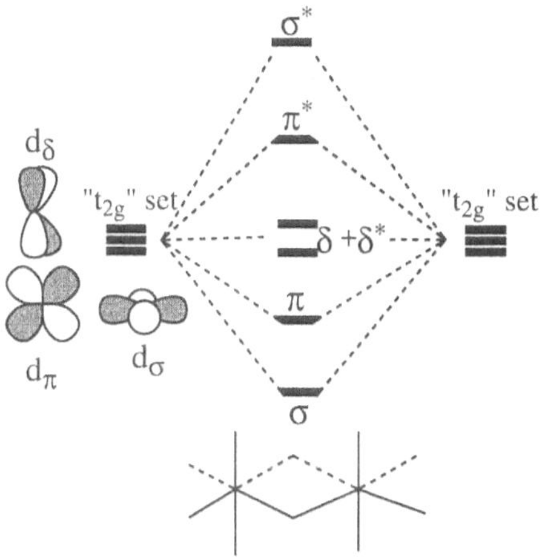

Fig. 3. Orbital interaction diagram for edge-sharing bioctahedral clusters

metal d electron counts. From the table, ESBOs of the early transition metals usually have total d electron counts of 2 through 6, while those of the later transition metals have d electron counts of 8 through 12. The optimal electron counts of four and eight, representing a formally M = M double bond, are relatively uncommon.

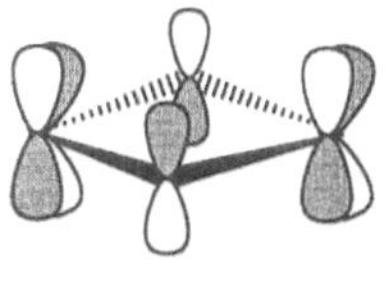

1

The bridging π-donor ligands play a dominant role in metal-metal bonding, as seen by the significant change of M-M bond length by, for example, replacement of a bridging Cl by Br. It is also noted that ESBO clusters with odd number of electrons are very rare.

3.3 Dinuclear Clusters Based on Face-Sharing Bioctahedra

Like edge-sharing bioctahedral clusters, each metal center in dinuclear clusters based on face-sharing bioctahedra (FSBO) has locally an octahedral ligand environment. The metal-metal orbital interaction can be derived from the two "t_{2g}" sets of the two metal centers. The orbital interaction diagram is shown in Fig. 4, having three bonding (a_1+e) and three antibonding (a_1* + e*) molecular orbitals. Here, we use the symmetry labels of a C_{3v} point group instead of the pseudo linear representations, (σ, π and δ used in previous sections). The reason is that the e orbitals contain both π and δ bonding characters though the a_1 orbital is almost pure σ bonding. In this case, the "t_{2g}" orbitals of each metal center defined with respect to the C_3 axis are: $d_{z2}(a_1)$, $(2/3)^{1/2}d_{x2-y2}+(1/3)^{1/2}d_{xz}$, and $(2/3)^{1/2}d_{xy}-(1/3)^{1/2}d_{yz}$ (e). From Fig. 4, one concludes that the optimal electronic structure has the three bonding orbitals fully occupied. Typical examples of FSBO clusters are listed in Table 1. Although the optimal metal d electron count is six, corresponding to the filling of all bonding orbitals, a similar variety of electron counts is found as the edge-bridging clusters. However, the later transition metal FSBO clusters usually have a metal d electron count of 12, resulting in no net metal-metal bonding. This is again a manifestation of the dominant role of bridging ligands in bringing the metal centers together.

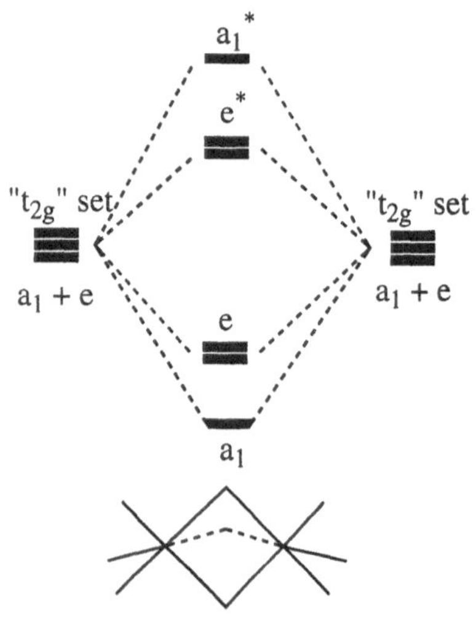

Fig. 4. Orbital interaction diagram for face-sharing bioctahedral clusters

3.4
Dinuclear Clusters with Quadruple Bridging Ligands

Two typical structural types exist for dinuclear clusters with quadruple bridging ligands. The first one normally contains η^5-cyclopentadienyl and η^6-arene (or its derivatives) terminal ligands (see Fig. 5). For clusters of this structural type, the ligand environment around each metal center has been commonly described as a "four-legged piano-stool" unit, and is electronically equivalent to seven-coordination. For this formally 7-coordinate metal fragment, only two d orbitals (see Fig. 5) are available for metal bonding because one of the "t_{2g}" set of orbitals in an octahedral complex is used to form an additional metal-ligand σ bond. The metal-metal orbital interaction, shown in Fig. 5, is therefore mainly derived from these available d orbitals, giving one σ bonding, $\delta + \delta^*$ roughly non-bonding, and one σ^* antibonding molecular orbitals. When the bridging ligands have available π donating orbitals, the δ orbital is expected to be higher in energy than the δ^* orbital. Examples with metal d electron counts of four and six are listed in Table 1.

Another type of quadruply bridged clusters consists of a square-antiprismatic structural unit for each metal center. With an additional metal orbital used for metal-ligand σ bonding, only one d_z2 orbital (the z axis is defined along the C_4 rotational axis) is available for metal-metal bonding. The metal-metal orbital interaction between the two d_z2 orbitals in this structural type of cluster gives rise to one σ bonding and one σ^* antibonding molecular orbital. Clusters with two metal d electrons provides the optimal electron count with an M-M single bond (see Table 1).

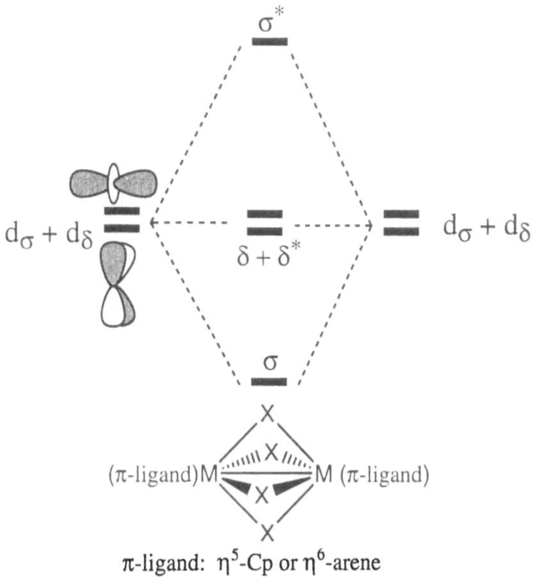

Fig. 5. Orbital interaction diagram for (η^6-arene)M(μ_2-X)$_4$M(η^6-arene) dinuclear clusters

3.5
L_3MML_3 Dinuclear Clusters

This type of unbridged dinuclear complexes constitutes an important class of metal alkoxide clusters (see Table 1). Clusters belonging to this class are formed mainly by Mo(III) and W(III), with a metal-metal triple bond. Each metal center uses six metal hybridized orbitals to form three metal-ligand and three metal-metal bonds, leaving three (hybridized) orbitals available for additional metal-ligand interaction. Therefore the ligands for clusters with this structural type are normally strong π-donor ligands, such as alkoxide (–OR) and amide ($-NR_2$). Usually, clusters containing $-NR_2$ ligands have a planar structure around each nitrogen center, indicating a strong metal-NR_2 interaction. The detailed electronic structures of various L_3MML_3 clusters have been reported in the literature [107]. The majority of clusters with local ML_3 units have an electronic configuration of d^3-d^3, i.e., a total of six metal d electrons (see Table 1).

3.6
Other Mo-Mo or W-W Dinuclear Clusters

The structural type shown in **2** is worth mentioning here as there are many such complexes characterized to date. These clusters normally contain the $M_2E_4^{2+}$ (M = Mo or W) structural unit with two M = E terminal double bonds (see 2). Usually, each metal center in these clusters can be formally assigned with a d^1 electronic configuration. These clusters can be viewed as having one metal-metal bond with a typical bond length of 2.85 Å. Reviews on these clusters can be found in the literature [108].

M=Mo and W; E=O, S and Se

2

4
Trinuclear Metal Clusters

Transition metal clusters with a triangular arrangement of metal centers are of fundamental interest since most of the observed molecular metal clusters have deltahedral skeletons. The structural variety brought about by capping (μ_3) and/or bridging (μ_2) ligands provides opportunities for novel ligand transformations, which may model the catalytic reactions occurring on the surface of larger clusters. The $Mo_3(\mu_2\text{-}S)_3$ unit, in particular, is closely related to the biologically important metal-sulfur cluster enzymes. The triangular cluster unit also finds great synthetic utility in building up larger cluster structures through the "template" approach [109]. There is a recent review by Saito on a wide variety of trinuclear clusters [109].

The elucidation of the electronic structure of trinuclear transition metal clusters has attracted continued theoretical studies, and for some types of triangular clusters the electronic structure has been studied by EHMO calculations [110]. The seminal review by Hoffman and co-workers [110] built up molecular orbital energy levels by successive addition of capping, bridging, and terminal ligands to the M_3 linear combinations of d orbitals. In this section we focus on five structure types which can be described by a local ML_n, n=5, 6, 7, using the same approach we have developed for the various types of dinuclear clusters above.

4.1
$M_3(\mu_2\text{-}X)_3L_9$

A typical example of this type is the classic cluster $Re_3(\mu_2\text{-}Cl)_3Cl_9^{3-}$ [18], which led to the recognition of metal-metal multiple bonds. Cotton and his co-worker have shown that there are six doubly occupied molecular orbitals along the three Re-Re edges, giving each edge double bonds. The cluster can be viewed as consisting of three ML_5 fragments, each having four frontier fragment orbitals (hy(σ) and the "t_{2g}" set (see Fig. 1)) available for metal-metal bonding. The linear combinations of three hy(σ)s span $a_1'+e'$ while the three "t_{2g}" sets give rise to combinations of $a_2''+e''$, $a_1''+e''$, and $a_2'+e'$ (see the left column in Fig. 6). The result of orbital interaction among these 12 linear combinations gives 6 bonding ($a_1'+a_2''+e'+e''$) and 6 antibonding ($a_1''+a_2'+e'+e''$) molecular orbitals, shown in Fig. 6. The occupation of the six bonding molecular orbitals

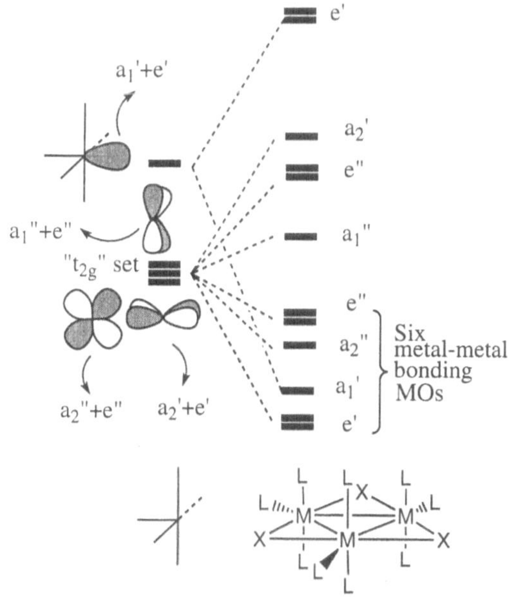

Fig. 6. Orbital interaction diagram for $M_3(\mu_2\text{-}X)_3L_{12}$ clusters

gives the optimal electronic structure with each M-M bond having a bond order of two. The rhenium cluster provides an example of this optimal electronic count of 12.

4.2 $M_3(\mu_3\text{-}E)(\mu_2\text{-}X)_3L_9$

This structural type provides an example of monocapped trinuclear clusters. Each metal center has locally an octahedral ligand environment, and therefore only the "t_{2g}" orbitals remain available. The linear combinations of the three sets of "t_{2g}" fragment orbitals transform as a_1+e, a_2+e, and a_1+e (see the left column in Fig. 7). The interaction between these combinations results in a pattern of three bonding (a_1+e), three roughly non-bonding (a_1+e, slightly antibonding character), and three strongly antibonding (a_2+e) molecular orbitals. Table 2 lists some typical examples of this structural type, the majority of which have six metal d electrons, i.e., the three metal-metal bonding orbitals are all occupied. Due to the existence of the three non-bonding (or slightly antibonding) orbitals, metal d electrons from 7 through 12 are possible, although for early transition metals the electron count does not exceed 9. The slightly antibonding nature of the "nonbonding" orbitals is manifested by the variation of M-M bond distances of $[Mo_3(\mu_3\text{-}O)(\mu_2\text{-}Cl)_3(\mu_2\text{-acetato})_3Cl_3]^-$ [122] and $[Mo_3(\mu_3\text{-}O)(\mu_2\text{-}Cl)_3(\mu_2\text{-acetato})_3Cl_3]^{2-}$ [123], which differ by only an electron. The ordering of the three roughly non-bonding orbitals has been discussed elsewhere [121].

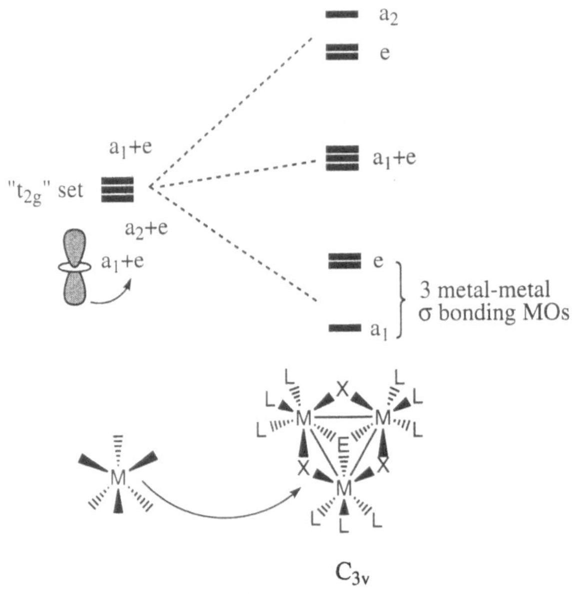

Fig. 7. Orbital interaction diagram for $M_3(\mu_3\text{-}E)(\mu_2\text{-}X)_3L_9$ clusters

Table 2. Representative examples of dinuclear transition metal clusters

Cluster Compound	Number of metal-metal bonding electrons	M-M bond distance (Å)	Reference
$M_3(\mu_2\text{-}X)_3L_9$			
$Re_2(\mu_2\text{-}Cl)_3Cl_6(PEt_2Ph)_3$	12	2.492	[111]
$[Re_2(\mu_2\text{-}Cl)_3Cl_9]^{3-}$	12	2.460–2.467	[112]
$M_3(\mu_3\text{-}E)(\mu_2\text{-}X)_3L_9$			
$Ti_3(\mu_3\text{-}CMe)(\mu_2\text{-}O)_3(\eta^5\text{-}C_5Me_5)_3$	0	2.819–2.824	[113]
$[Nb_3(\mu_3\text{-}S)(\mu_2\text{-}O)_3(NCS)_9]^{6-}$	4	2.760–2.770	[114]
$[Nb_3(\mu_3\text{-}Cl)(\mu_2\text{-}Cl)_3Cl_6(PEt_3)_3]^{1-}$	6	2.969–2.988	[115]
$Nb_3(\mu_3\text{-}Cl)(\mu_2\text{-}Cl)_3Cl_3(PMe_2Ph)_6$	8	2.824–2.836	[116]
$[Ta_3(\mu_3\text{-}Cl)(\mu_2\text{-}Cl)_3Cl_6(PEt_3)_3]^{1-}$	6	2.928–2.938	[115]
$Cr_3(\mu_3\text{-}CH)(\mu_2\text{-}Cl)_3(\eta^5\text{-}Cp)_3$	9	2.795–2.837	[117]
$[Mo_3(\mu_3\text{-}O)(\mu_2\text{-}O)_3(H_2O)_9]^{4+}$	6	2.471–2.489	[118]
$[Mo_3(\mu_3\text{-}S)(\mu_2\text{-}S)_3Cl_3(dmpe)_3]^{+}$	6	2.768	[119]
$Mo_3(\mu_3\text{-}As)(\mu_2\text{-}O)_3(\eta^5\text{-}Cp)_3$	6	2.664–2.667	[120]
$Mo_3(\mu_3\text{-}S)(\mu_2\text{-}S)_3(\eta^5\text{-}CpMe_5)_3$	7	2.854–2.865	[121]
$[Mo_3(\mu_3\text{-}O)(\mu_2\text{-}Cl)_3(\mu_2\text{-acetato})_3Cl_3]^{1-}$	8	2.566–2.572	[122]
$[Mo_3(\mu_3\text{-}O)(\mu_2\text{-}Cl)_3(\mu_2\text{-acetato})_3Cl_3]^{2-}$	9	2.598–2.617	[123]
$[W_3(\mu_3\text{-}S)(\mu_2\text{-}S)_3(NCS)_9]^{5-}$	6	2.764–2.767	[124]
$[W_3(\mu_3\text{-}O)(\mu_2\text{-}Cl)_3(\mu_2\text{-acetato})_3Cl_3]^{3-}$	8	2.566–2.568	[122]
$[W_3(\mu_3\text{-}O)(\mu_2\text{-}Cl)_3(\mu_2\text{-acetato})_3Cl_3]^{2-}$	9	2.577–2.612	[125]
$M_3(\mu_3\text{-}E)(\mu_2\text{-}X)_6L_6$			
$[Ti_3(\mu_3\text{-}O)(\mu_2{:}\eta^2\text{-}S_2)_3Cl_6]^{2-}$	0	3.14–3.18	[126]
$[V_3(\mu_3\text{-}S)(\mu_2{:}\eta^2\text{-}S_2)_3(bipy)_3]^{1+}$	6	2.751–2.769	[127]
$[Mo_3(\mu_3\text{-}S)(\mu_2{:}\eta^2\text{-}S_2)_3Cl_6]^{2-}$	6	2.757–2.759	[128]
$[Mo_3(\mu_3\text{-}S)(\mu_2{:}\eta^2\text{-}S_2)_3(\eta^2\text{-}S_2)_3]^{2-}$	6	2.723–2.728	[108]
$[Mo_3(\mu_3\text{-}S)(\mu_2{:}\eta^2\text{-}S_2)_3Br_6]^{2-}$	6	2.741–2.751	[129]
$[Mo_3(\mu_3\text{-}Te)(\mu_2{:}\eta^2\text{-}Te_2)_3(TeI_3)_3]^{+}$	6	2.851–2.860	[130]
$[W_3(\mu_3\text{-}S)(\mu_2{:}\eta^2\text{-}S_2)_3Br_6]^{2-}$	6	2.733–2.744	[131]
$M_3(\mu_3\text{-}E)_2L_9$			
$Ru_3(\mu_3\text{-}Cl)(\mu_3\text{-}S)(\eta^2\text{-}Cp)_3$	18	2.874–2.901	[132]
$[Ru_3(\mu_3\text{-}S)_2(\eta^6\text{-arene})_3]^{2+}$	18	2.763–2.796	[133]
$Co_3(\mu_3\text{-}CH)_2(\eta^5\text{-}Cp)_3$	18	2.437	[134]
$[Co_3(\mu_3\text{-}S)_2(\eta^5\text{-}CpMe)_3]^{2+}$	18	2.486–2.546	[135]
$[Co_3(\mu_3\text{-}S)_2(\eta^5\text{-}CpMe)_3]^{1+}$	19	2.486–2.875	[135]
$[Co_3(\mu_3\text{-}S)_2(\eta^5\text{-}Cp)_3]$	20	2.624–2.759	[136]
$Rh_3(\mu_3\text{-}CH)_2(\eta^5\text{-}Cp)_3$	18	2.636	[137]
$[Rh_3(\mu_3\text{-}S)_2(\eta^5\text{-}Cp)_3]^{2+}$	18	2.830	[138]
$[Ir_3(\mu_3\text{-}S)_2(\eta^5\text{-}C_5Me_5)_3]^{2+}$	18	2.816–2.821	[139]
$M_3(\mu_3\text{-}E)_2(\mu_2\text{-}X)_3L_6$			
$[Ti_3(\mu_3\text{-}Cl)_2(\mu_2\text{-}Cl)_3(en)_3]^{1+}$	6	2.761–2.770	[140]
$[V_3(\mu_3\text{-}Cl)_2(\mu_2\text{-}Cl)_3(en)_3]^{1+}$	9	3.134–3.155	[141]
$Cr_3(\mu_3\text{-}S)_2(\mu_2\text{-}S)_3(dmpe)_3$	8	2.554–2.646	[142]

Table 2 (Continued)

Cluster Compound	Number of metal-metal bonding electrons	M-M bond distance (Å)	Reference
$Mo_3(\mu_3\text{-}O)(\mu_3\text{-}OR)(\mu_2\text{-}OR)_3(OR)_6(R=CH_2{}^tBu)$	6	2.523–2.539	[143]
$Mo_3(\mu_3\text{-}S)_2(\mu_2\text{-}S)_3(PMe_3)_6$	8	2.714	[144]
$M_3(\mu_3\text{-}E)_2(\mu_2\text{-}X)_6L_3$			
$[Nb_3(\mu_3\text{-}O)_2(\mu_2\text{-}butyrato)_6(thf)_3]^{1+}$	4	2.834–2.855	[145]
$[Mo_3(\mu_3\text{-}O)_2(\mu_2\text{-}acetato)_6(H_2O)_3]^{2+}$	6	2.749–2.755	[146]
$[W_3(\mu_3\text{-}O)_2(\mu_2\text{-}acetato)_6(H_2O)_3]^{2+}$	6	2.746	[147]

4.3 $M_3(\mu_3\text{-}E)(\mu_2\text{-}X)_6L_6$ and $M_3(\mu_3\text{-}E)_2L_9$

This structural type provides another example of monocapped trinuclear clusters. Each metal center has a local ML_7 ligand environment with two fragment orbitals available for metal-metal interaction. The pattern of the relevant metal-metal orbital spectrum is shown in Fig. 8. The optimal electron count is six, i.e., all bonding orbitals are occupied. Some typical examples are listed in Table 2 together with their metal d electron counts and metal-metal bond distances. Many of these clusters have six metal d electrons because no nonbonding orbital is available.

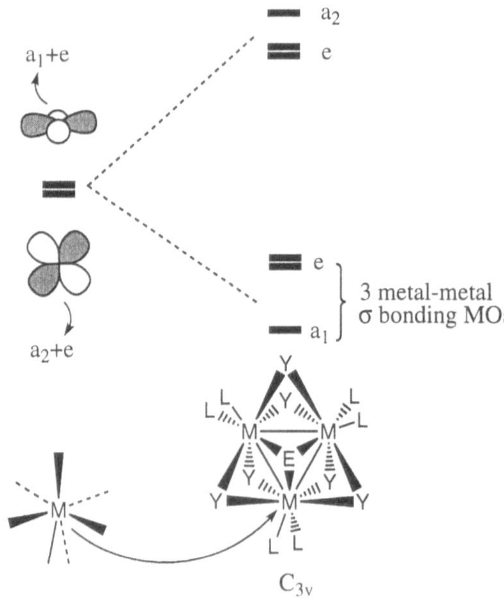

Fig. 8. Orbital interaction diagram for $M_3(\mu_3\text{-}E)(\mu_2\text{-}X)_6L_6$ clusters

$M_3(\mu_3\text{-}E)_2L_9$ clusters listed in Table 2 are atypical among transition halide clusters in that they contain late transition metals. When ligand electron pairs are included, these clusters have an electron count of around 48, which is the normal electron count for a triangle carbonyl cluster. The detailed electronic structures will not be elaborated here.

4.4 $M_3(\mu_3\text{-}E)_2(\mu_2\text{-}X)_3L_6$

This is a bicapped trinuclear structural type. Each metal center has an octahedral ligand environment. The linear combinations of the three sets of "t_{2g}" fragment orbitals span $a_1'+e'$, $a_2'+e'$, and $a_1''+e''$ (see the left-hand side of Fig. 9). As a result of the orbital interaction among these linear combinations, three metal-metal bonding ($a_1'+e'$), three slightly antibonding ($a_1''+e''$), and three strongly antibonding ($a_2'+e'$) molecular orbitals are generated. The slightly antibonding character of $a_1''+e''$ is a result of their interaction with the π-donor orbitals of the three μ_2-X bridging ligands. The orbital pattern is very similar to the case described in Sect. 4.2 (see Fig. 7). The optimal electron count for these clusters is again six, with electron counts up to nine caused by the filling of the slightly antibonding orbitals. Some typical

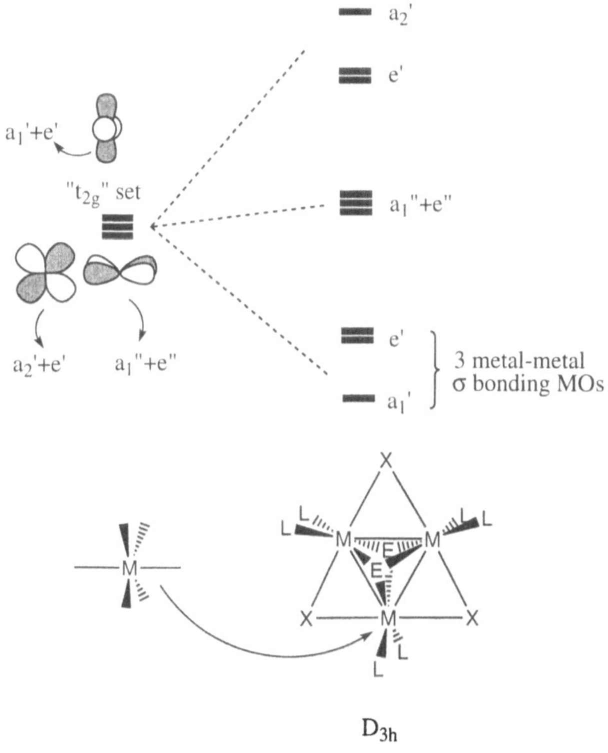

Fig. 9. Orbital interaction diagram for $M_3(\mu_3\text{-}E)_2(\mu_2\text{-}X)_3L_6$ clusters

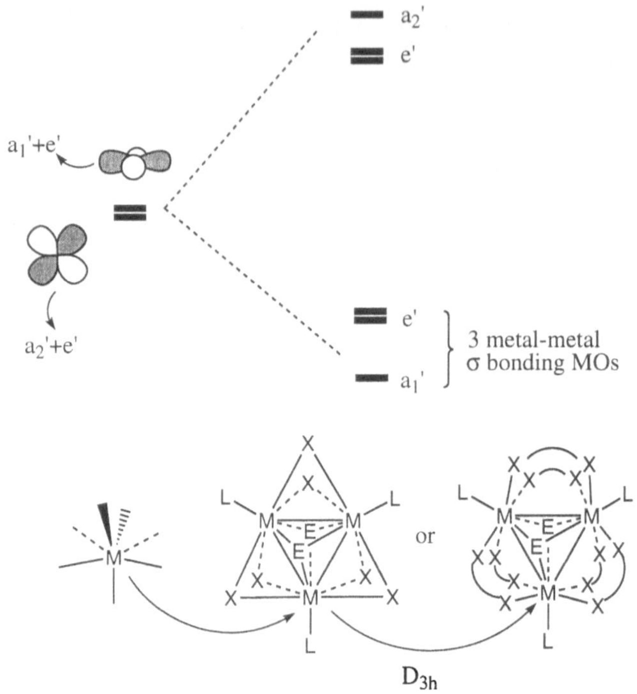

Fig. 10. Orbital interaction diagram for $M_3(\mu_3\text{-}E)_2(\mu_2\text{-}X)_6L_3$ clusters

examples with this structural type together with their electron counts are listed in Table 2.

4.5
$M_3(\mu_3\text{-}E)_2(\mu_2\text{-}X)_6L_3$

This is another type of bicapped trinuclear cluster. The metal fragment has an ML_7 local environment, thus having two metal d orbitals available for metal-metal bonding. These two metal d orbitals are approximately non-bonding for an ML_7 complex as shown in Fig. 10. The result of orbital interaction among the three sets of these two available d orbitals leads to three strongly bonding ($a_1'+e'$) and three strongly antibonding ($a_2'+e'$) metal-metal molecular orbitals (see Fig. 10). Unlike the cases discussed in Sects. 4.2 and 4.4, clusters with this structural type normally do not have more than six metal-metal bonding electrons because any additional electron would have to occupy the strongly antibonding orbitals. Table 2 lists some typical examples which contain six or fewer metal d electrons.

5
Tetranuclear Metal Clusters

The structural variety of clusters increases rapidly with nuclearity. For tetranuclear clusters three types of regular structures can be perceived: tetrahedral, edge-shared triangular ("butterfly" or rhombus), and square planar (square or rectangle), corresponding to the successive opening of the compact *closo* structure. Various structural types of tetranuclear clusters containing alkoxyl ligands have been reviewed [148]. In tetranuclear carbonyl clusters, the structure opening through bond breaking has been shown as a result of the addition of skeletal electrons [6]. The situation is not so simple for halide clusters, due to the possibility of M-M multiple bonding and delocalized metal orbital interactions. There have been, to date, few systematic structural and bonding studies of tetranuclear clusters with π-donor ligands, and we will refrain from providing details of metal frontier orbital interaction. Nevertheless, the structural variety and electron counts shown below will provide insights to the interesting bonding characteristics within these clusters.

5.1 Square-Planar Clusters

Square-planar structure is one of the three typical structures for tetranuclear clusters. The two niobium clusters listed in Table 3 represent examples with a regular structure having eight metal-metal bonding electrons. In other words, there are four equivalent formally single metal-metal bonds in these two clusters (see Fig. 11a). The 12-electron cluster $Mo_4(\mu_2\text{-O-}^iPr)_8Cl_4$ [151] adopts a structure shown in Fig. 11b with four equivalent metal-metal bonds. Another cluster with a similar structure $[Mo_4(\mu_2\text{-}Cl_8)Cl_4]^{3-}$ has 15 electrons, and adopts a rectangular structure somewhat related to the well-known distortion of cyclobutadiene. Apparently, further molecular orbital calculations are needed in order to understand the bonding peculiarities in these two clusters. The cluster $Mo_4(\mu_2\text{-}Cl)_4Cl_4(PMe_3)_4$ has a rectangular structure shown in Fig. 11c. The metal-metal bonding has been described as having four alternate Mo_2 triple and single bond units (see Fig. 11d) [159], which requires 16 metal d electrons. As expected from the dominant role of bridging ligands in binding metal centers, the Mo-Mo bond with bridging ligands is much longer than the unbridged one.

In Table 3, several late transition metal clusters having a square-planar structure are also listed. In these clusters, metal-metal bonding interaction becomes weak as late transition metals have poor d orbital overlap due to their contracted size. Consequently, some metal-metal antibonding molecular orbitals are occupied, as discussed above for dinuclear clusters. Therefore these late transition metal clusters have more d electrons when compared to the corresponding early transition metal ones.

5.2 Butterfly or Rhombic Clusters

Butterfly or rhombic clusters represent another type of tetranuclear cluster. Table 3 lists some typical examples. The first two clusters listed in the table adopt a rhombic structure shown in Fig. 11e. Both $Nb_4(\mu_3\text{-}Cl)_2(\mu_2\text{-}Cl)_4Cl_4(PMe_3)_6$ [160] and $Mo_4(\mu_3\text{-}S)_2(\mu_2\text{-}S)_4(PMe_3)_6(SH)_2$ [161] have ten metal-metal bonding electrons, implying that each edge of the rhombic structure corresponds to one metal-metal bond. $Mo_4(\mu_3\text{-O}^iPr)_2(\mu_2\text{-O}^iPr)_4(O^iPr)_2Br_4$ [162]

Table 3. Selected examples of tetranuclear transition metal clusters

Cluster Compound	Number of metal-metal bonding electrons	M-M bond distance (Å)	Reference
Square-Planar Structures			
$[Nb_4(\mu_4\text{-}S)_2(\mu_2\text{-}SPh)_8(SPh_4]^{4-}$	8	2.826–2.828	[149]
$[Nb_4(\mu_4\text{-}S)_2(\mu_2\text{-}SPh)_8(PMe_4Ph]_4$	8	2.810–2.822	[150]
$Mo_4(\mu_2\text{-}O^iPr)_8Cl_4$	12	2.407	[151]
$Mo_4(\mu_2\text{-}Cl)_4Cl_4(PMe_3)_4$	16	2.226, 2.878	[152]
$Mo_4(\mu_2\text{-}Cl)_8Cl_4]^{3-}$	15	2.355, 2.654	[153]
$Co_4(\mu_4\text{-}S)_4Cp_4$	28	2.438–2.439	[154]
$Ni_4(\mu_2\text{-}S\text{-cyclohexenyl})_8$	32	2.681–2.692	[155]
$Ni_4(\mu_4\text{-}Se)_2Cp_4$	32	2.570–2.580	[156]
$Pd_4(\mu_2\text{-}OOC\text{-}CCl_3)_4(\mu_2\text{-}OO^tBu)_4$	32	2.903–2.944	[157]
$Pt_4(\mu_2\text{-acetato})_8$	32	2.487–2.501	[158]
Butterfly or Rhombic Structures			
$Nb_4(\mu_3\text{-}Cl)_2(\mu_2\text{-}Cl)_4Cl_4(PMe_3)_6$	10	2.92 × 4, 2.985	[160]
$Mo_4(\mu_3\text{-}S)_2(\mu_2\text{-}S)_4(PMe_3)_6(SH)_2$	10	2.83 × 4, 2.845	[161]
$Mo_4(\mu_3\text{-}O^iPr)_2(\mu_2\text{-}O^iPr)_4(O^iPr)_2Br_4$	12	2.51 × 4, 2.481	[162]
$[Mo_4(\mu_4\text{-}O)(\mu_3\text{-}Br)_2(\mu_2\text{-}Br)_4Br_6]^{2-}$	12	2.67 × 4, 2.593	[163]
$W_4(\mu_2\text{-}O^iPr)_4O^iPr)_8$	12	2.50 × 2, 2.73 × 2, 2.807	[164]
$[Mo_4(\mu_3\text{-}O)(\mu_2\text{-}OCH_2{}^tBu)_4(OCH_2{}^tBu)_7]^-$	12	2.48 × 2, 2.61 × 2, 2.422	[165]
$W_4(\mu_3\text{-}SOEt)_2(\mu_2\text{-}OEt)_4(OEt)_{10}$	8	2.94 × 2, 2.65 × 2, 2.763	[166]
$W_4(\mu_4\text{-}C)(\mu_2\text{-}NH_2Me)(\mu_2\text{-}OPr)_4(OPr)_8$	8	2.74–2.82	[167]
Tetrahedral Structures			
$V_4(\mu_3\text{-}S)_4(\eta^5\text{-}MeCp)_4$	8	2.867–2.883	[169]
$[V_4(\mu_3\text{-}S)_4(\eta^5\text{-}MeCp)_4]^+$	7	2.852–2.855	[169]
$Cr_4(\mu_3\text{-}S)_4(\eta^5\text{-}MeCp)_4$	12	2.822–2.848	[170]
$[Mo_4(\mu_3\text{-}S)_4(\eta^5\text{-}{}^iPrCp)_4]^{2+}$	10	2.790–2.897	[171]
$[Mo_4(\mu_3\text{-}S)_4(\eta^5\text{-}{}^iPrCp)_4]^+$	11	2.860–2.923	[171]
$Mo_4(\mu_3\text{-}S)_4(\eta^5\text{-}{}^iPrCp)_4$	12	2.892–2.912	[171]
$[Mo_4(\mu_3\text{-}S)_4(CN)_{12}]^{8-}$	12	2.855	[172]
$Re_4(\mu_4\text{-}Se)_4Te_{12}$ unit in $Re_4Se_4Te_4$ solid	12	2.785	[173]

adopts a butterfly structure, shown in Fig. 11f. For those clusters having more or less than ten electrons, the corresponding butterfly structure experiences a significant distortion except for $Mo_4(\mu_3\text{-}O^iPr)_2(\mu_2\text{-}O^iPr)_4(O^iPr)_2Br_4$ [162] and $[Mo_4(\mu_4\text{-}O)(\mu_3\text{-}Br)_2(\mu_2\text{-}Br)_4Br_6]^{2-}$ [163] (see Table 3). The electronic and steric factors determining the degree of bending of the butterfly "wings", and the role of μ_4 ligands in "tying" the wings, would be of interest.

5.3 Tetrahedral Clusters

Clusters belonging to this type normally have four μ_3-S or μ_3-Se ligands. These clusters are also known as cubane-type clusters since the arrangements of

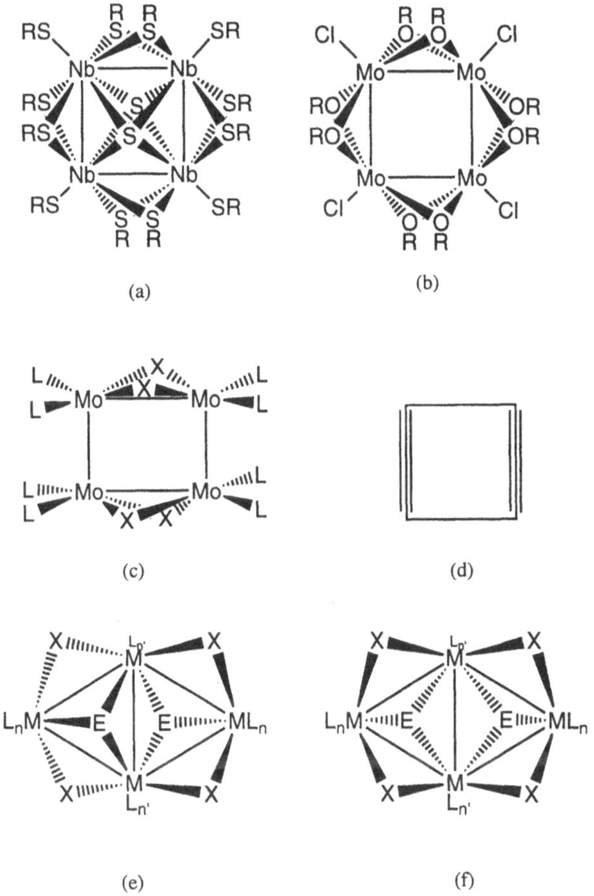

Fig. 11. Some typical square-planar, butterfly and rhombic tetranuclear clusters

metal and chalcogen atoms form approximately a cube. Tetrahedral clusters normally possess six metal-metal bonds. Therefore, one would expect that most of the tetrahedral clusters have 12 metal-metal bonding d electrons. The relevant molecular orbital interaction diagram has been provided in the literature [168] and will not be discussed in more detail here.

Most of the examples have cyclopentadienyl or arene terminal ligands, resulting in a local "three-legged piano-stool" coordination. In Table 3, clusters with 10 and 11 electron counts also exist. The $V_4(\mu_3\text{-}S)_4(\eta^5\text{-}MeCp)_4$ [169] cluster has only eight metal-metal bonding d electrons. This is not too surprising in view of the fact that a tetrahedral cluster can have only four skeletal electron pairs, such as B_4Cl_4.

6
Pentanuclear Metal Clusters

Two typical geometries for pentanuclear metal clusters are square-pyramidal and trigonal-bipyramidal. $Zr_5(\mu_3\text{-}H)_2(\mu_2\text{-}H)_2(\mu_2\text{-}Cl)_8Cl_4(PMe_3)_5$ [174] provides an example of the square-pyramidal structure. In this cluster, each edge is bridged by one μ_2-Cl ligand and each metal has one terminal phosphine ligand. The four basal metal atoms have additional Cl terminal ligands. The cluster can be seen as an *arachno* structure of an edge-bridged octahedral cluster and has 12 skeletal bonding electrons. Here, the Hs are considered to provide electrons to the skeletal bonding. For an edge-bridged octahedral cluster, eight skeletal bonding electron pairs are the optimal electron count (see discussion in the next section), which corresponds to eight localized electron pairs in the eight triangle faces of the octahedron. For the Zr_5 cluster, one could suggest that the square face accommodates an extra electron pairs. A more detailed electronic structure analysis is required in order to have a better picture.

$V_5(\mu_3\text{-}O)_6(\eta^5\text{-}Cp)_5$ [178], $[V_5(\mu_3\text{-}S)_6(\eta^5\text{-}CpMe)_5]^+$ [179], and $Ti_5(\mu_3\text{-}E)_6(\eta^5\text{-}Cp')_5$ (E = Se or S, Cp′= Cp or CpMe) [175] have a trigonal-bipyramidal structure. In these clusters, each metal atom has one terminal Cp ligand and the six μ_3-E (E = O or Se) ligands cap the six triangle faces. Further examination of the structure reveals that the two apical metal atoms have locally a three-legged

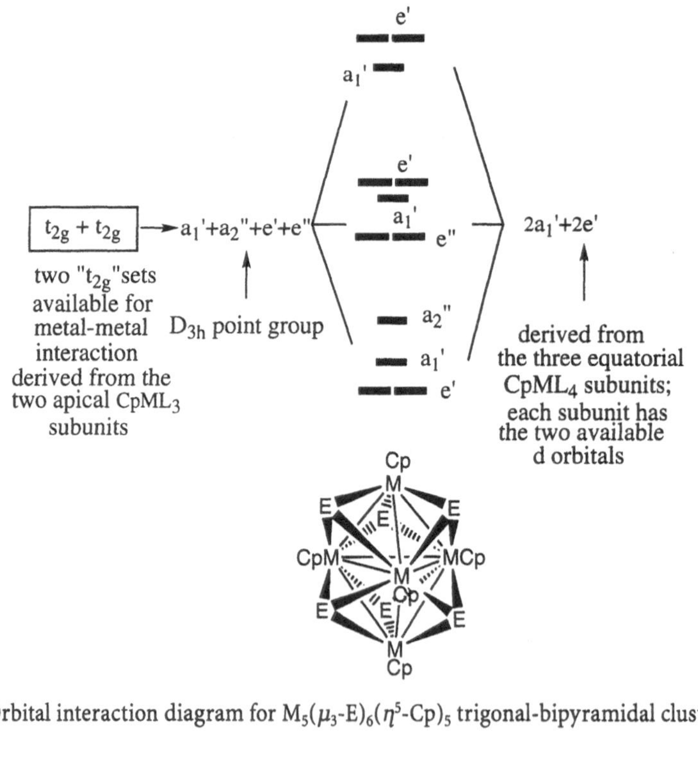

Fig. 12. Orbital interaction diagram for $M_5(\mu_3\text{-}E)_6(\eta^5\text{-}Cp)_5$ trigonal-bipyramidal clusters

Table 4. Trigonal-bipyramidal transition metal clusters

Formula	Metal d Electron count	M-M bond distance (Å) M_{eq}-M_{eq}	M_{eq}-M_{ax}	Reference
$Ti_5(\mu_3\text{-}S)_6(\eta^5\text{-}Cp)_5$	3	3.15–3.21	3.08–3.17	[175]
$Ti_5(\mu_3\text{-}Se)_6(\eta^5\text{-}Cp)_5$	3	3.37–3.38	3.32	[176]
$Ti_5(\mu_3\text{-}Se)_6(\eta^5\text{-}CpMe)_5$	3	3.33–3.39	3.30–3.35	[177]
$V_5(\mu_3\text{-}O)_6(\eta^5\text{-}Cp)_5$	8	2.74	2.74–2.75	[178]
$[V_5(\mu_3\text{-}S)_6(\eta^5\text{-}CpMe)_5]^+$	7	3.21–3.25	2.97–3.01	[179]

piano-stool ligand environment while the three equatorial metal atoms have locally a four-legged piano-stool ligand environment. A three-legged piano-stool complex can be regarded as a pseudo octahedral molecule and therefore has "t_{2g}" orbitals available for metal-metal interaction. For a four-legged piano-stool complex, two d orbitals are available. Based on the orbital interaction concept, one can schematically construct the relevant metal-metal orbital interaction diagram, shown in Fig. 12. $V_5(\mu_3\text{-}O)_6(\eta^5\text{-}Cp)_5$ has eight metal d orbitals, corresponding to the occupation of the four low-lying metal-metal bonding molecular orbitals ($e' + a_1' + a_2''$) in the figure. $[V_5(\mu_3\text{-}S)_6(\eta^5\text{-}CpMe)_5]^+$ has seven metal d electrons and provides an example of partial occupation of the four bonding molecular orbitals. For the titanium clusters, only three metal d electrons are available for metal-metal bonding. Molecular orbital calculations by Bottomley and Day [176] showed that these three d electrons occupy the e′ orbitals (see Fig. 12). Table 4 summarizes the relevant metal-metal bond distances for these clusters together with their metal d electron counts. A similar table has been given by Bottomley and Day [176].

7 Hexanuclear Metal Clusters

7.1 Octahedral Clusters

Face- and edge-bridged (**3**) structures are the two typical types of octahedral clusters containing π-donor ligands. They have the general formula of $M_6(\mu_3\text{-}X)_8L_6$ and $M_6(\mu_2\text{-}X)_{12}L_6$ (M: transition metal; X: bridging ligand; and L: terminal ligand). Some clusters contain an interstitial atom, either main group or transition metal, which can be formulated as $M_6(\mu_3\text{-}X)_8(\mu_6\text{-}E)L_6$ and $M_6(\mu_2\text{-}X)_{12}(\mu_6\text{-}E)L_6$, where E is the interstitial atom. Due to their structural frequency and unusual physical properties, these octahedral clusters have attracted considerable experimental and theoretical studies [180–185]. Recently we developed a unified picture in understanding their structure and bonding based on the orbital interaction idea [16].

In the unified molecular orbital picture [16], the metal-metal orbital interactions are derived from the combinations of the frontier orbitals of six square-pyramidal ML_5 fragments since each transition metal in these octahedral

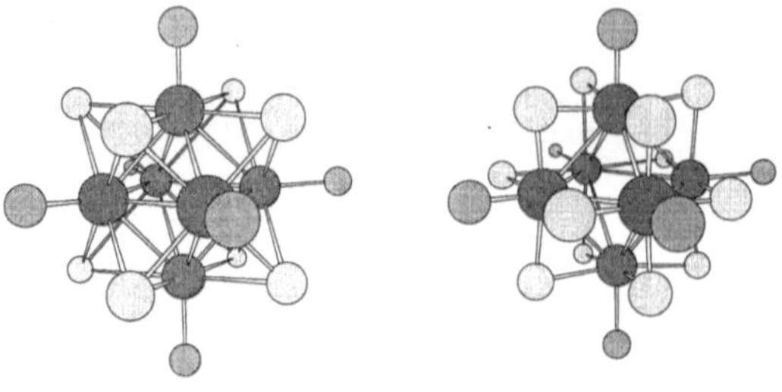

3

clusters has locally five metal-ligand σ bonds. The frontier orbitals of an ML_5 fragment are a σ-bonding hybrid, hy(σ), and the "t_{2g}" set (see Fig. 1). The hy(σ) is derived from one hybridized orbital of an ML_6 octahedral complex. The three "t_{2g}" orbitals of each ML_5 fragment can be divided into a d(π) pair and one d(δ) component based on the number of nodes with respect to a radial vector (from the centroid of a given cluster to a metal atom). A d(π) component has one node and the d(δ) has two nodes.

The molecular orbitals derived from the interactions between the hy(σ) and d(π) pairs of the six ML_5 fragments should resemble the skeletal orbital interactions of the *closo*-$B_6H_6^{2-}$ cluster in which each BH fragment also has one hy(σ) and one pair of π frontier orbitals. For this *closo* octahedral cluster, there are 7 skeletal bonding orbitals ($a_{1g}+t_{1u}+t_{2g}$) and 11 antibonding orbitals ($t_{1g}+t_{2u}+e_g{}^*+t_{1u}{}^*$). The energy ordering of these 18 molecular orbitals is more or less the same for both face- and edge-bridged octahedral clusters (see Fig. 13c,e).

The interactions among the d(δ) basis sets depend greatly on the arrangement of ligands, which leads to the different characteristic electron counts for the two types of octahedral clusters. In Fig. 13c, e, a face-bridged cluster $M_6(\mu_3\text{-}X)_8L_6$ has five bonding molecular orbitals ($t_{2u}+e_g$) and one antibonding molecular orbital (a_{2g}) whereas an edge-bridged cluster $M_6(\mu_2\text{-}X)_{12}L_6$ has one bonding molecular orbital (a_{2u}) and five antibonding molecular orbitals (e_u+t_{2g}).

In summary, $M_6(\mu_3\text{-}X)_8L_6$ (face-bridged) has 12 M-M bonding molecular orbitals, which can be unitarily transformed into 12 localized orbitals having the maximum amplitude along the 12 M-M edges. For $M_6(\mu_2\text{-}X)_{12}L_6$ (edge-bridged), there are eight M-M bonding molecular orbitals, which can be unitarily transformed into eight localized orbitals having the maximum amplitude on the eight triangle faces. The full occupation of these M-M bonding orbitals leads to a stable electron count of 24 metal electrons for a face-bridged cluster (12 metal-metal 2-center-2-electron bonds) and 16 for an edge-bridged cluster (8 metal-metal 3-center-2-electron bonds). One can appreciate the two different „magic" numbers as follows. When the ligands are face-bridged, the metal electrons occupy the regions along the M-M bonds to avoid electron-electron repulsion. Similarly, when the ligands are edge-bridged, the metal electrons move to the regions of metal triangle faces.

When an interstitial main group atom is present, the interaction molecular orbital diagrams for these situations are illustrated in Fig. 13b,f. As can be seen, the

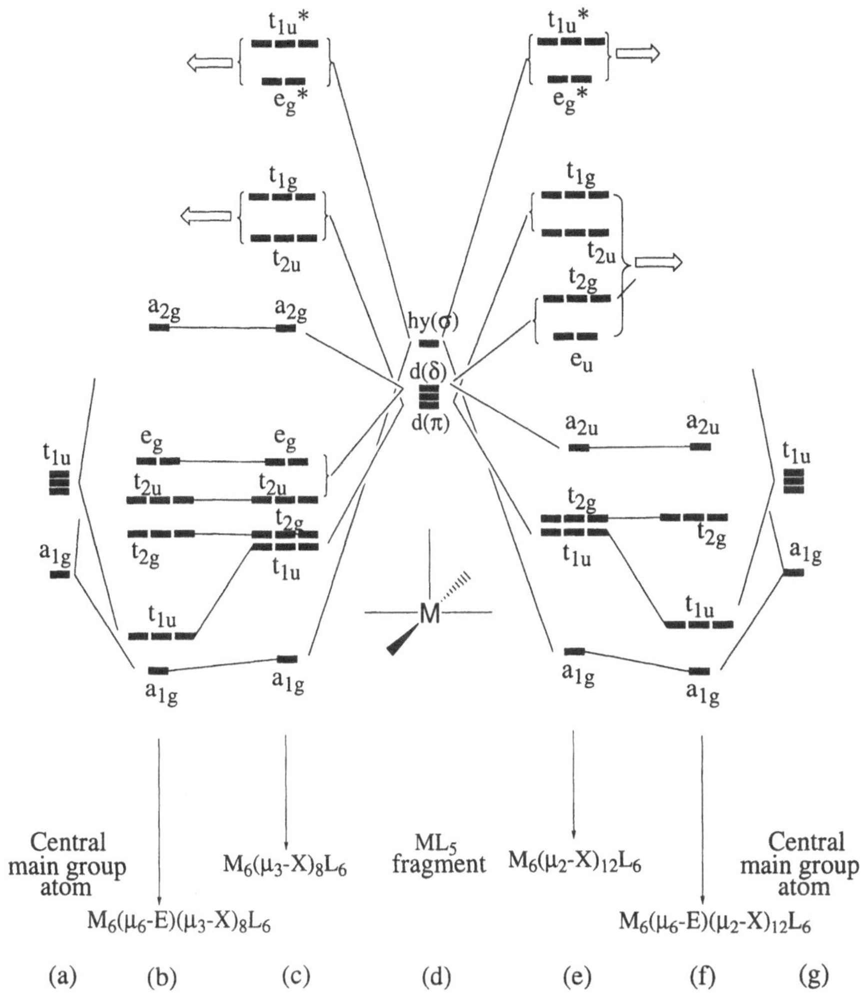

Fig. 13. Orbital interaction diagrams for face- and edge-bridged octahedral clusters (Adapted from: Lin Z and Williams ID (1996) Polyhedron 15:3277)

total number of bonding molecular orbitals remains unchanged for both types of octahedral cluster. The net effect is that the a_{1g} and t_{1u} levels are stabilized.

When an interstitial transition metal atom is present, the situation becomes slightly different. Orbital interaction diagrams for these cluster types are shown in Fig. 14. For the face-bridged case, the total number of bonding molecular orbitals remains unchanged at 12 but the orbital ordering is slightly altered. The t_{2g} and e_g orbitals are stabilized through their bonding interactions with the d orbitals of the interstitial metal atom (Fig. 14a). For the edge-bridged case, two d orbitals (e_g) of the central metal cannot match symmetry-adapted orbitals from the parent cluster and become non-bonding (see Fig. 14b). Therefore, the total number of metal-metal and metal-central atom bonding molecular orbitals becomes ten. As shown in Fig. 14, electron counts of 18 (9 lowest molecular

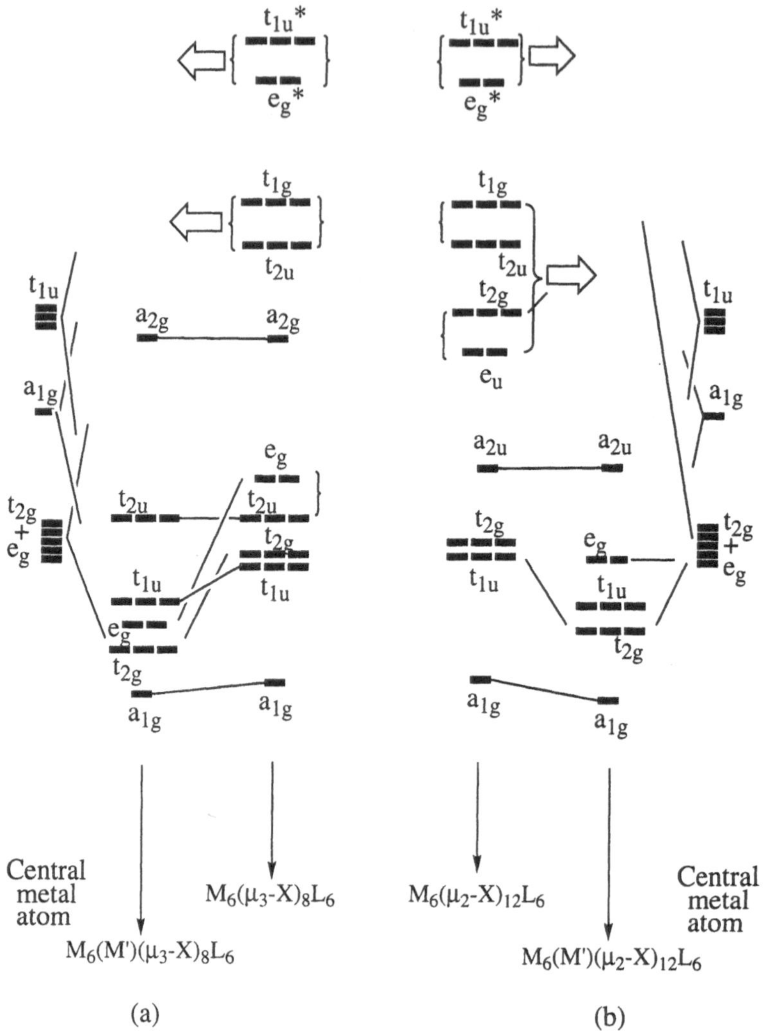

Fig. 14. Orbital interaction diagrams for face- and edge-bridged octahedral clusters with an interstitial transition-metal atom (Adapted from: Lin Z and Williams ID (1996) Polyhedron 15:3277)

orbitals are occupied) also give significant HOMO-LUMO gaps for both the face- and edge-bridged clusters.

In summary, those clusters with an interstitial main group atom should have the same closed-shell electronic requirements as their parent structures (see Fig. 13b,f). The interstitial main group atom contributes its valence electrons to the metal-metal and metal-interstitial atom bonding. For a given octahedral cluster with an interstitial main group atom, the electron count includes the valence electrons of the interstitial atom. For clusters with an interstitial transition metal atom, the closed-shell requirements depend on whether the cluster is face-bridged or edge-bridged. In view of the orbital interaction diagram

(Fig. 14), however, one can see that both face- and edge-bridged clusters with an interstitial transition metal can have an electron count of 18 only when the 9 lowest orbitals are fully occupied (Fig. 14a,b), leaving the t_{2u} for face-bridged clusters and a_{2u} for edge-bridged clusters unoccupied, respectively.

Some typical examples of face- and edge-bridged octahedral clusters are listed in Table 5 together with their metal skeletal electron counts and metal-metal distances. For face-bridged clusters, the majority of early transition metal clusters listed under this class indeed conform to the electron count of 24 discussed above. There are also a number of clusters having an electron count of 20. These clusters leave the e_g orbitals unoccupied (see Fig. 13(c)). The Fe and Co clusters listed in Table 5 possess up to 38 metal d electrons. Since the d orbitals of Fe and Co are relatively contracted there is less direct metal-metal bonding, and the energy splitting due to their metal-metal interaction is relatively small. It is expected that the seven antibonding molecular orbitals ($a_{2g}+t_{2u}+t_{1g}$ of Fig. 13c) derived from the d(π) and d(δ) basis sets for these Fe and Co clusters are only slightly antibonding. The occupation of these seven additional molecular orbitals will not severely destabilize the cluster, thus leading to a maximum number of 38 valence electrons. The Fe clusters in Table 5 are examples of partial occupation of the seven additional molecular orbitals, and display paramagnetic properties. It has been shown that the complex $[Fe_6(\mu_3\text{-}S)_8(PEt_3)_6]^+$ [212] possesses a $S=7/2$ spin state with an electronic configuration of $[(a_{1g})^2(t_{1u})^6(t_{2g})^6(t_{2u})^6(e_g)^4(a_{2g})^1(t_{2u})^3(t_{1g})^3]$. In these Fe and Co clusters, clusters with fewer valence electrons tend to have shorter metal-metal bond distances because there is less occupation of the antibonding molecular orbitals.

For edge-bridged clusters, although the closed-shell electron count for an edge-bridged octahedral cluster is 16, the majority of such clusters listed in Table 5 have electron counts of 14, 15, or 16. Apparently, an electron count of 14 leaves the a_{2u} orbital vacant (see Fig. 13e) in which a significant HOMO-LUMO gap can also be attained. Several 15-electron clusters half-fill the a_{2u} molecular orbital.

Several structurally characterized examples of edge-bridged octahedral clusters with an interstitial transition metal atom are known (see Table 5). These clusters have 18 (or 17) metal electrons with an electronic configuration of $(a_{1g})^2(t_{2g})^6(t_{1u})^6(e_g)^4$ (see Fig. 14b). This electronic configuration implies that the interstitial transition metal atom meets the 18-electron rule and that the interactions between the interstitial transition metal atom and outer ones dominate the metal-metal bonding characteristics.

Some clusters with electron counts deviated from the ideal case above have been discussed in detail in our previous paper [16], where we also discussed clusters with η^5-Cp terminal ligands. Due to space constraint, the details will not be elaborated here.

7.2 Trigonal Prismatic Clusters

There are not many examples of trigonal prismatic clusters containing π-donor ligands to be found in the literature, the earliest examples being synthesized by

Table 5. Some representative octahedral clusters

Formula	Electron count[a]	M-M bond distance (Å)	Reference
Face-Bridged Octahedral Clusters			
Clusters without an interstitial atom			
$[Mo_6(\mu_3\text{-}S)_8(PEt_3)_6]$	20	2.66	[186]
$[Mo_6(\mu_3\text{-}Se)_8(PEt_3)_6]$	20	2.705	[187]
$[Mo_6(\mu_3\text{-}Cl)_8)Cl_6]^{2-}$	24	2.59–2.61	[188]
$Pb[Mo_6(\mu_3\text{-}I)_8I_6]$	24	2.67	[189]
$[W_6(\mu_3\text{-}S)_8(PEt_3)_6]$	20	2.68	[190]
$[W_6(\mu_3\text{-}Cl)_8Cl_4(PBu^n_3)_2]$	24	2.60–2.62	[191]
$[Re_6(\mu_3\text{-}S)_4(\mu_3\text{-}Cl)_4Cl_6]$	24	2.59–2.60	[192]
$[Re_6(\mu_3\text{-}Se)_5(\mu_3\text{-}Cl)_3Cl_6]^-$	24	2.61–2.62	[193]
$[Fe_6(\mu_3\text{-}S)_8(PEt_3)_6]^{2+}$	30	2.62	[194]
$[Fe_6(\mu_3\text{-}Te)_8(PMe_3)_6]$	32	2.82–2.97	[195]
$[Co_6(\mu_3\text{-}S)_8(PR_3)_6]^+$	37	2.77–2.84	[196]
$[Co_6(\mu_3\text{-}S)_8(CO)_6]$	38	2.80–2.82	[197]
$[Co_6(\mu_3\text{-}Te)_8(PEt_3)_6]$	38	3.225	[198]
Clusters with an interstitial main group atom			
$V_6(\mu_3\text{-}Se)_8(\mu_6\text{-}O)(PMe_3)_6$	20	2.80–2.84	[199]
Edge-Bridged Octahedral Clusters			
Clusters without an interstitial atom			
$[Nb_6(\mu_2\text{-}Cl)_{12}Cl_6]^{2-}$	14	3.01–3.02	[200]
$[Nb_6(\mu_2\text{-}Cl)_{12}Cl_6]^{3-}$	15	2.94–2.99	[201]
$[Nb_6(\mu_2\text{-}Br)_{12}Br_6]^{4-}$	16	2.97–2.98	[202]
$[Ta_6(\mu_2\text{-}Cl)_{12}(OMe)_6]^{2-}$	14	2.98–2.99	[203]
$[Ta_6(\mu_2\text{-}Cl)_{12}(MeOH)_6]^{3+}$	15	2.89–2.91	[204]
$[Ta_6(\mu_2\text{-}Cl)_{12}(CN)_6]^{4-}$	16	2.931	[205]
Clusters with an interstitial atom (main group)			
$Sc[Sc_6(\mu_2\text{-}Br)_{12}(\mu_6\text{-}C)]$	13	3.21–3.30	[206]
$[Ti_6(\mu_2\text{-}Cl)_{12}(\mu_6\text{-}C)Cl_6]^{4-}$	14	3.00–3.10	[207]
$[Zr_6(\mu_2\text{-}Cl)_{12}(\mu_6\text{-}C)Cl_6]^{4-}$	14	3.22–3.24	[208–209]
$[Ti_6(\mu_6\text{-}Be)Br_{15}]^{5-}$	16	3.35–3.39	[210]
Clusters with a transition metal intestitial atom			
$Sc[Sc_6(\mu_2\text{-}I)_{12}(\mu_6\text{-}Co)]$	18	3.39–3.49	[211]
$Y[Y_6(\mu_2\text{-}I)_{12}(\mu_6\text{-}Fe)]$	17	3.67–3.75	[211]
$[Zr_6(\mu_2\text{-}Cl)_{12}(\mu_6\text{-}Fe)Cl_6]^{4-}$	18	3.38–3.39	[209]

[a] Number of valence electrons involving metal-metal and metal-interstitial bonding.

Russians [213–216]. Some typical examples of technetium and rhenium clusters are shown in Fig. 15. For bonding comparison, some octanuclear Tc clusters are also included in the figure. From Fig. 15, the Tc-Tc (or Re-Re) distances are remarkably short. The Tc-Tc (or Re-Re) distances on the triangle faces are much shorter than those parallel to the prism axis. $Tc_6Cl_{12}^{2-}$ and $Tc_6Cl_{14}^{3-}$ [213] have 32 and 31 metal d electrons, respectively while Re_6Br_{12} [215]

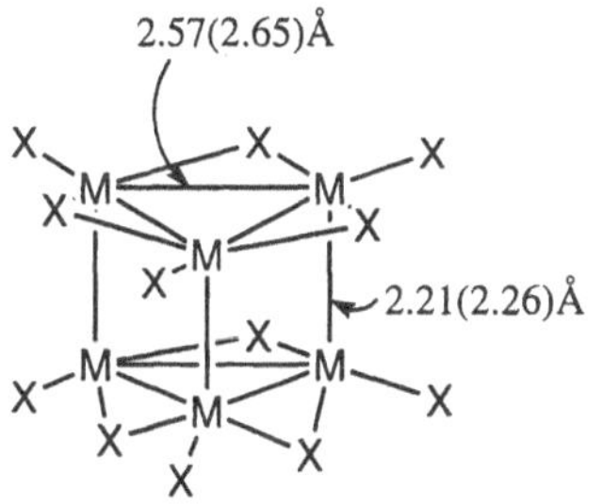

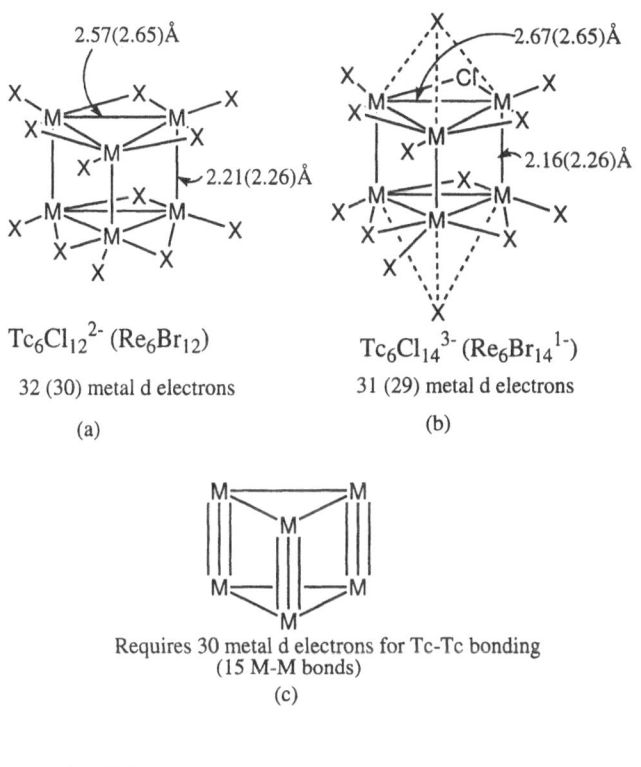

$Tc_6Cl_{12}^{2-}$ (Re_6Br_{12})

32 (30) metal d electrons

(a)

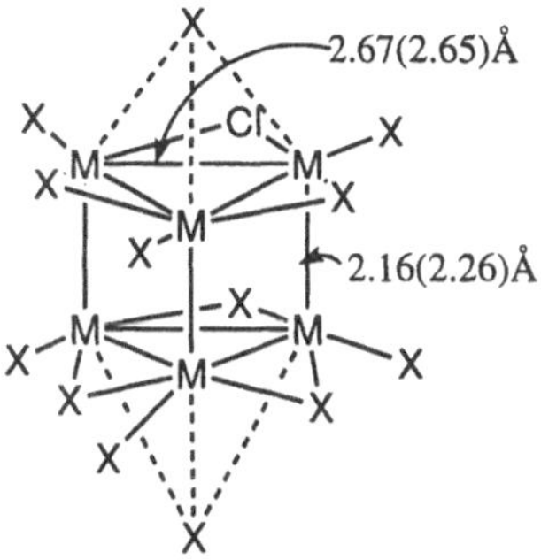

$Tc_6Cl_{14}^{3-}$ ($Re_6Br_{14}^{1-}$)

31 (29) metal d electrons

(b)

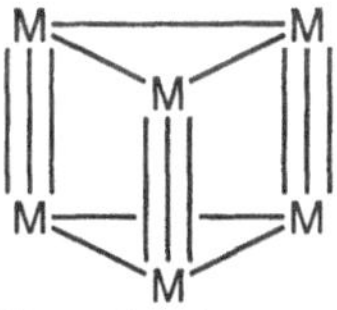

Requires 30 metal d electrons for Tc-Tc bonding (15 M-M bonds)

(c)

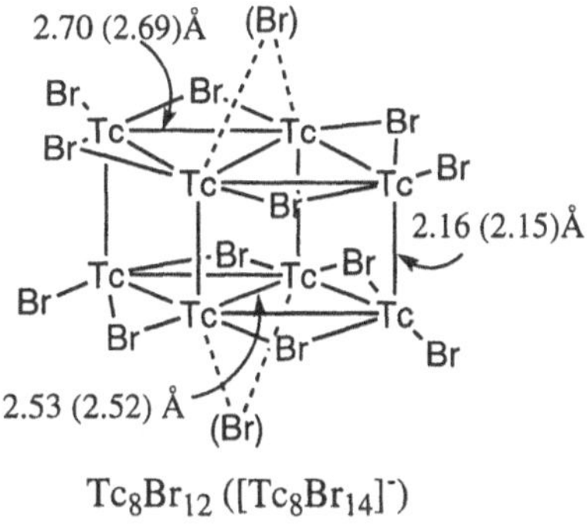

Tc_8Br_{12} ($[Tc_8Br_{14}]^-$)

44 (43) metal d electrons

(d)

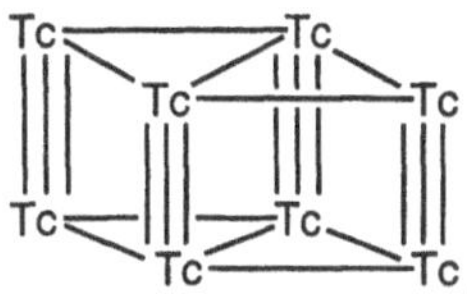

Requires 44 metal d electrons for Tc-Tc bonding (22 M-M bonds)

(e)

Fig. 15. Some typical trigonal-prismatic clusters

and $Re_6Br_{14}^{1-}$ [216] have 30 and 29. Detailed molecular orbital analyses on their metal-metal bonding by Wheeler and Hoffmann [159] gave single bonds in the triangles and multiple bonding between triangles. Figure 15c schematically illustrates the corresponding bonding picture, in which 30 metal d electrons are required in order to form the 15 metal-metal bonds. The extra two electrons of $Tc_6Cl_{12}^{2-}$ occupy an a_2'' molecular orbital that is weakly bonding within the triangles, but antibonding between the two triangles.

The bonding picture (Fig. 15c) can be appreciated as follows. Each M uses s, p_x, p_y, $d_{xy,}$ and d_{x2-y2} metal orbitals (z axis is defined along the prism axis) to form five σ bonds with three Cl and two other Ms within the triangle. The d_{z2}, d_{xz} and d_{yz} orbitals, which have their maximum amplitudes between the two triangles, can be utilized to form triple bonds along the prism edge with the technetium from the other triangle. The unused p_z orbital of each Tc can be used to attract additional electron donors as shown in the cluster $Tc_6Cl_{14}^{3-}$ (or $Re_6Br_{14}^{1-}$) (see Fig. 15b).

Tc_8Br_{12} and $Tc_8Br_{14}^{-}$ [214] have 44 and 43 metal electrons, respectively. The metal-metal interaction in these two clusters is more or less similar to that for the two hexanuclear technetium clusters. The bonding picture is schematically illustrated in Fig. 15e in which 44 metal electrons are required to form 22 metal-metal bonds. The arguments above can be used to appreciate the detailed metal-metal interaction.

Another trigonal-prismatic cluster, $Nb_6SBr_{18}^{4-}$ [217], was reported more recently. This cluster can be formulated as $[Nb_6(\mu_6\text{-}S)(\mu_2\text{-}Br)_{12}Br_6]^{4-}$. In the prismatic structure, each triangle edge has one μ_2-Br bridging ligand while each edge along the prismatic axis has doubly bridged (i.e., two μ_2-Br bridging ligands). The Nb-Nb distances along the edges of rectangles are 3.28 Å, which are much longer than those along the edges of the two triangles (2.97 Å). In view of the difference between the two distinct types of Nb-Nb bond distances, the cluster has been described as two Nb_3 triangles held together by strong Nb-S bonds, and the direct interaction between the two triangles is quite weak. Formally, the cluster has 14 metal d electrons. Molecular orbital calculations [217] show that 12 metal d electrons fill the Nb-Nb bonding orbitals, which form the two triangles. The remaining two electrons occupy the weakly Nb-Nb bonding orbital with respect to the three edges along the prism axis.

Although the factors leading to the preference of the trigonal prism to the more usual octahedron for all these clusters are still not fully understood, it is generally believed that the more electron-rich clusters (in this case >28) require a more open structure, as manifested by the presence of rectangles in these trigonal prismatic structures.

8 Higher Nuclearity Metal Clusters

Recently, it was reported [218] that the crystal structure of $PrMo_8O_{14}$ contains both *cis* and *trans* bi-face-capped octahedral Mo_8 clusters. In these clusters, the oxygens act as bridging and terminal ligands not only within individual cluster but also between clusters in the solid state. The octahedral unit in these two clusters can be regarded as an edge-bridged octahedral structure with μ_2-O bridging ligands. Around those capping Mo atoms, the Os are, however, μ_3 bridging. These two bi-capped Mo_8 clusters have 24 metal d electrons if Pr^{4+} is assumed. A localized bonding picture may be proposed here for these clusters, i.e., the 12 electron pairs are approximately localized at the six unbridged triangle faces from the central octahedral unit and the six unbridged edges from the capping units. In addition to the two bi-capped octahedral geometries discussed above, cube is another important type of octanuclear clusters [219]. Some of the cubic clusters will be covered in the following two sections. Others are discussed in another chapter of this book by Halet and his co-workers.

More recently, McCarley and co-workers [220] reported a series of $M_3Mo_{14}O_{22}$ solid state structures (M_3=K_3, $K_{1.66}Pb_{1.34}$ and $K_{1.29}Sn_{1.71}$). These solids contain oligomeric cluster units with three *trans* edged-shared Mo octahedra in their solid states. Each octahedral subunit is again μ_2-O edge-bridged. The formal d electron counts for these oligomeric clusters are approximately 44. Examining the structural details of these clusters, there are 16 unbridged triangle faces and 4 unbridged edges. If a localized bonding picture is applicable, one would expect that the formal d electron count is 40, which is slightly less than the observed electron count of 44. The four extra electrons may occupy two Mo-Mo antibonding molecular orbitals, resulting in the significant lengthening of a couple of Mo-Mo bonds. The majority of Mo-Mo bond distances in these clusters are in the range of 2.75–2.80 Å while a couple of them have longer Mo-Mo bond lengths of around 3.05 Å.

9
Mo-Fe-S and Related Metal Clusters

Due to their biological importance, the Mo-Fe-S and related clusters have attracted considerable interest. For example, the active site of nitrogenase consists of Mo-Fe-S clusters [221]. During the last two decades, a large number of such clusters have been synthesized and characterized, and there is now an abundance of excellent reviews on their syntheses and structures [222–225]. These clusters, in general, can be classified into two broad categories: (1) chain clusters containing $M(\mu_2\text{-}S)_2Fe$ unit (M=Mo or W); and (2) cubane-like clusters containing $MFe_3(\mu_3\text{-}S)_4$ (M=Mo, W, or Re). Some typical examples of these clusters are illustrated in Figs. 16 and 17.

Although there have been several quantum-mechanical studies of these clusters [226], the relevant metal-metal interaction is still not very clear to date. In these clusters, metal-metal interaction is believed to occur through the so-called spin-spin coupling and is considered to be very weak. This hypothesis is supported by the fact that each Fe center retains its high-spin nature in the clusters. The discussions on the electronic structures for these clusters are normally focused on their metal oxidation states and electron spin properties. Due to the weak metal-metal interaction, a well-defined closed-shell electronic structure, i.e., a general electron counting rule, for these clusters may not be possible.

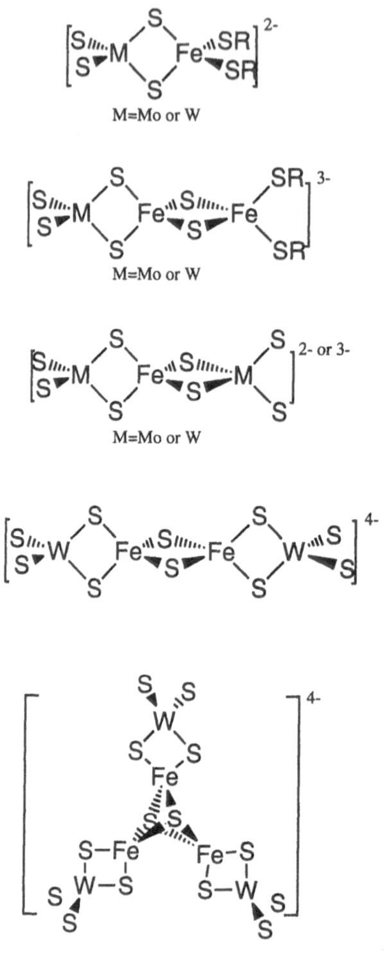

Fig. 16. Various structural types of Mo-Fe-S chain-type clusters

9.1 Chain-Like Clusters

$[S_2M(\mu_2\text{-}S)_2Fe(SR)_2]^{2-}$ (M=Mo or W) [223–224] provides the simplest example of this type. The Mössbauer chemical shift of ^{57}Fe shows that the oxidation state of Fe is somewhere between high-spin Fe^{2+} and Fe^{3+}. Mo or W would have a formal oxidation state of +6 if Fe has +2. The result of the Mössbauer experiment implies that there is a charge transfer from Fe^{2+} to Mo^{6+} (or W^{6+}), and the extreme case of complete electron transfer would imply that Mo (or W) has an oxidation state of +5. An orbital interaction model, shown in Fig. 18a, can be used to explain the charge transfer and the relevant spin multiplicity. In the figure, S_2ML_2 is considered to be isolobal with $CpML_2$, and has one available d_{z2} orbital (M-Fe axis is defined along the z axis) for M-Fe bonding while $Fe(SR)_4$'s five metal d orbitals split into $e+t_2$ under a pseudo local T point group. The

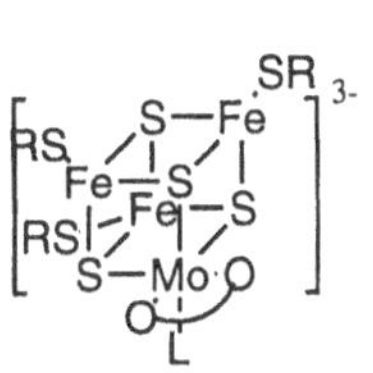

$\{MoFe_3\}^{11+}$ 19 d electrons

(a)

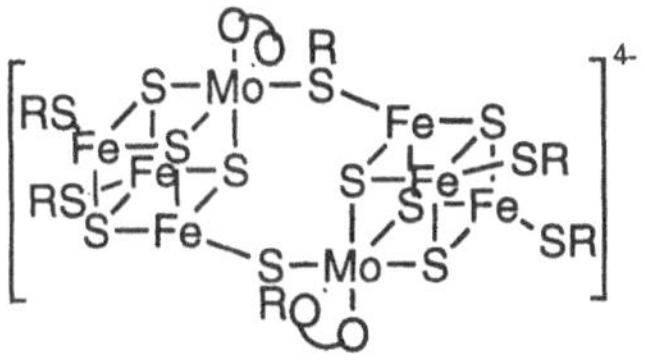

$\{Mo_2Fe_6\}^{22+}$ 38 d electrons

(b)

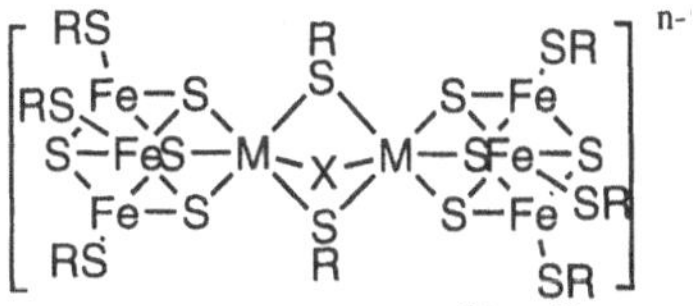

(a) M=Mo and X=SR (n=3 or 5) $\{Mo_2Fe_6\}^{22+ \text{ or } 20+}$ 38 or 40 d electrons

(b) M=Mo and X=S (n=3) $\{Mo_2Fe_6\}^{23+}$ 39 d electrons

(c) M=Re and X=SR (n=3) $\{Re_2Fe_6\}^{22+}$ 40 d electrons

(c)

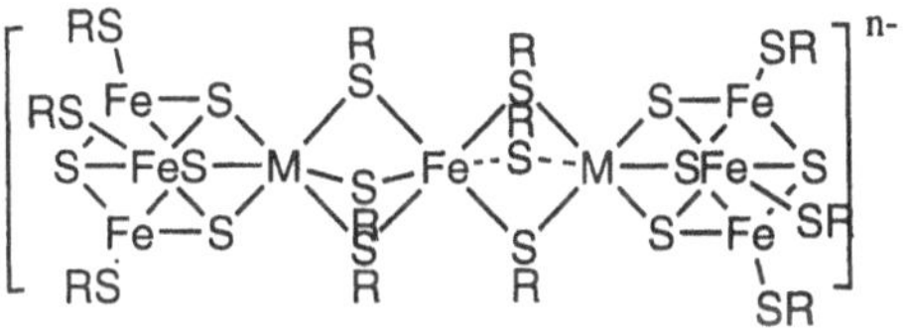

(a) M=Mo (n=3 or 4) $\{MoFe_3\}^{11+} + Fe^{3+ \text{ or } 2+} + \{MoFe_3\}^{11+}$

(b) M=Re (n=4) $\{ReFe_3\}^{11+} + Fe^{3+} + \{ReFe_3\}^{11+}$

(d)

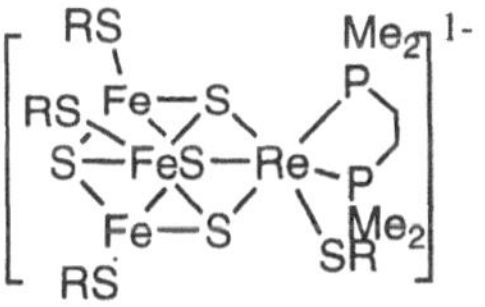

$\{ReFe_3\}^{11+}$ 20 d electrons

(e)

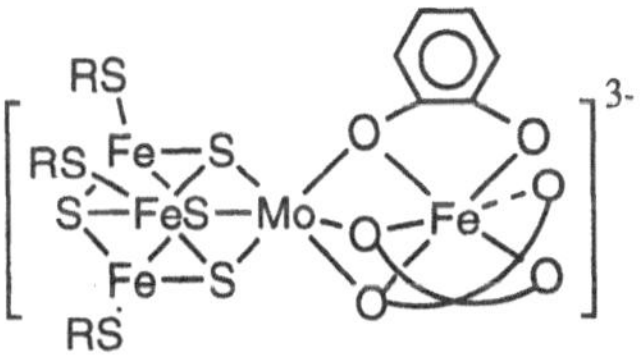

$\{MoFe_3\}^{11+} + Fe^{3+}$

(f)

Fig. 17. Various structural types of Mo-Fe-S cubane-type clusters

$M(d_{z^2})$-$Fe(d_{z^2})$ weak bonding orbital, which has two electrons contributed mainly from Fe, is responsible for the observed charge transfer. Due to the small splitting of the five Fe d orbitals, the cluster has a spin multiplicity of five (four unpaired d electrons).

Because of the $M(d_{z^2})$-$Fe(d_{z^2})$ bonding interaction, Fe has a formal oxidation of +2, i.e., formally a high-spin d^6 electronic configuration with the paired electrons as a weak metal electron donor being able to interact with the empty d_{z^2} orbital of the S_2ML_2 fragment. This $M(d_{z^2})$-$Fe(d_{z^2})$ bonding may be very

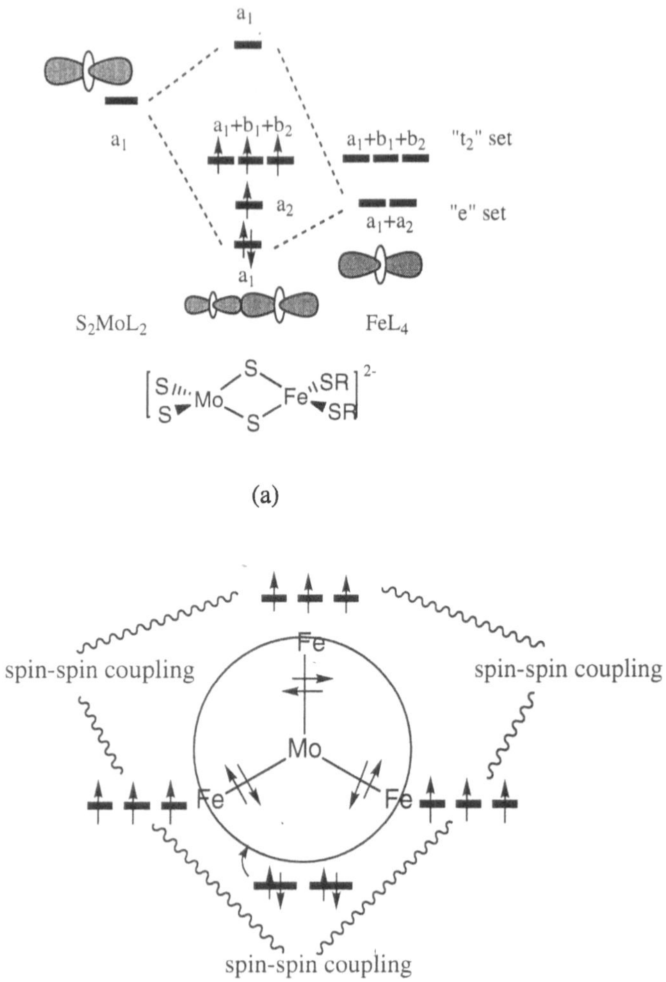

Metal-Metal Bonding Model for $\{MoFe_3S_4\}^{3+}$

(b)

Fig. 18. (a) Metal d-d orbital interactions in Mo-Fe chain clusters; (b) Bonding model for $\{MoFe_3S_4\}^{3+}$ cubane-type clusters

important for all chain-like clusters. Carefully examining the formal oxidation states for these chain-like clusters (see Fig. 16), one finds that the formal oxidation states for the iron bonded directly to M (Mo or W) are all +2. However, $[S_2Mo(\mu_2\text{-}S)_2Fe(\mu_2\text{-}S)_2MoS_2]^{3-}$ [223] has a formal Fe oxidation state of +1, if one assumes an oxidation state of +6 for Mo. For those iron atoms not directly bonded to Mo or W, their oxidation states can be formally +2 or +3.

Fe-Fe interaction in these clusters is normally described as a spin-coupling interaction. For example, the doublet spin structure (S = 1/2) of $[S_2Mo(\mu_2\text{-}S)_2$

$Fe(\mu_2\text{-}S)_2Fe(SPh)_2]^{3-}$ [227] has been explained in terms of the spin-coupling between two high-spin centers, Fe(+2), bonded directly to the Mo, and Fe(+3). For $[Cl_2Fe(\mu_2\text{-}S)_2Mo(\mu_2\text{-}S)_2FeCl_2]^{2-}$ [228], its diamagnetic spin state, S=0, has been explained with an antiferromagnetic coupling of the two non-interacting high-spin Fe(+2) centers through the two weak Mo-Fe bonds.

9.2
Cubane-Like Clusters

Both single- and double-cubane clusters have been synthesized and characterized. Various examples are shown in Fig. 17. The physico-chemical properties of these clusters have been studied in detail using Mössbauer and electron paramagnetic resonance experiments [222], showing that the oxidation states of Mo and Fe are +3 and ca. +2.67, respectively in most cases. In general, double-cubane clusters consist of two single-cubane units linked by S or SR bridging ligands. The distance between the two single-cubane units in a double-cubane cluster is usually over 3.5 Å. Therefore, the electronic structure and magnetic properties of the double-cubane clusters closely resemble those of single-cubane clusters. The two types of cubane clusters are thus expected to have similar bonding characteristics.

For a single-cubane cluster, magnetic susceptibility measurements show that S=3/2. From the molecular formula of a single-cubane cluster, one can have $[MoFe_3]^{+11}$. This implies that each cubane has 19 metal d electrons. A metal-metal bonding interaction model (Fig. 18b) has been proposed to account for the physico-chemical properties described above through a detailed molecular orbital analysis based on the extended Hückel molecular orbital calculations [226c,d]. The bonding model (Fig. 18b) indicates that there are three weak Mo-Fe bonds, accommodating six metal d electrons, in a single-cubane cluster. Each Mo-Fe bond can be described as a Fe→Mo dative bond, similar to the one described above for the $Mo(\mu_2\text{-}S)_2Fe$ unit of chain-like clusters. In addition, two weak Fe-Fe bonds exist, accommodating four metal d electrons. The remaining nine metal d electrons fill in the three "t_2" sets derived from each FeL_4 fragment. The spin-spin coupling of these three "t_2" sets leads to an overall spin state of S=3/2. The existence of three weak Fe→Mo bonds explains the experimental observation of the Mo^{3+} formal oxidation state.

Figure 17 also illustrates the formal oxidation states for various single- and double-cubane clusters. It can be seen from the figure that the metal d electron count is approximately 19 for each single-cubane unit. In the bridging unit of the cluster in Fig. 17b, Mo-Fe is 4.38 Å. The Mo-Mo (or Re-Re) distances in Fig. 17c are over 3.30 Å. The Mo-Fe (or Re-Fe) (Fe here is the central iron) distances are also over 3.30 Å. The Mo-Fe distance in Fig. 17f is 3.092 Å. The assignments of the oxidation states for the distinct Fe in Fig. 17d, f are supported by Mössbauer Fe chemical shift measurements. More recently, $[MoFe_3S_4(SC_6H_{11})_3Fe(SC_6H_{11})_4]^{2-}$ [225] was synthesized and structurally characterized. The metal-sulfur framework of this cluster is similar to Fig. 17f except that the distinct Fe is four-coordinate instead of six-coordinate. Again, one can formulate this cluster as $\{MoFe_3\}^{11+} + \{Fe\}^{2+}$. These experimental observations imply that the double-cubane clusters have indeed similar bonding characteristics to the single-cubane clusters, and the relevant electron count can simply be taken as the sum of two single-cubane units. The interaction between the two single-cubane subunits for the double-cubane clusters should be very weak and antiferromagnetic.

9.3 Fe-S Clusters

Similar to the Mo-Fe-S clusters, Fe-S clusters do not have very well defined closed-shell electronic structures. The Fe-Fe interaction is again described as spin-spin coupling among the high-spin iron centers. Figure 19 illustrates some typical Fe-S clusters reported in the literature [229–235]. The average oxidation states are also given in the figure. From Fig. 19, various oxidation states in these Fe-S clusters are observed, and some clusters are stable in different oxidation states. Since the Fe-Fe interaction is quite weak in these Fe-S clusters, the average Fe oxidation states differ from one structure to another. Examining the structure-oxidation state correlation, one comes to the following two general conclusions:

(a) the more μ_2-S bridging ligands, the higher is the Fe average oxidation state;
(b) the more μ_3-S, μ_4-S, or μ_2-SR bridging ligands, the lower is the average Fe oxidation state.

The clusters in Fig. 19a (when E=S) and Fig. 19b contain only μ_2-S bridging ligands and have an average oxidation state of +3 (according to conclusion (a) above). The clusters in Fig. 19a (when E=SR) and Fig. 19c,f contain only μ_2-SR bridging ligands while the cluster in Fig. 19i has only μ_4-S bridging ligands. These clusters have the lowest Fe oxidation state (+2) (according to conclusion (b) above).

These observations can be understood as follows. In these Fe-S clusters, the Fe(d)-Fe(d) molecular orbitals, which the metal d electrons occupy, are expected to be destabilized by the available π-donating orbitals of sulfur. For a μ_2-S bridging ligand, there is one p_π orbital available for Fe-S π interaction. For μ_3-S, μ_4-S, or μ_2-SR bridging ligands, there is no p_π orbital available for Fe-S π interaction as all orbitals are utilized for Fe-S σ bonding. It is expected that the more available p_π orbitals from sulfur bridging ligands a cluster has, the more the Fe(d)-Fe(d) molecular orbitals of the cluster are destabilized. The result of such destabilization is that fewer metal d electrons are allowed, leading to a higher average Fe oxidation state.

10 Miscellaneous Metal Clusters

Some examples of nickel chalcogen (such as Ni-S and Ni-Se) clusters are worth mentioning. The chemistry of these nickel clusters has attracted much interest due to the unique biological function of nickel-sulfur sites in enzymes with

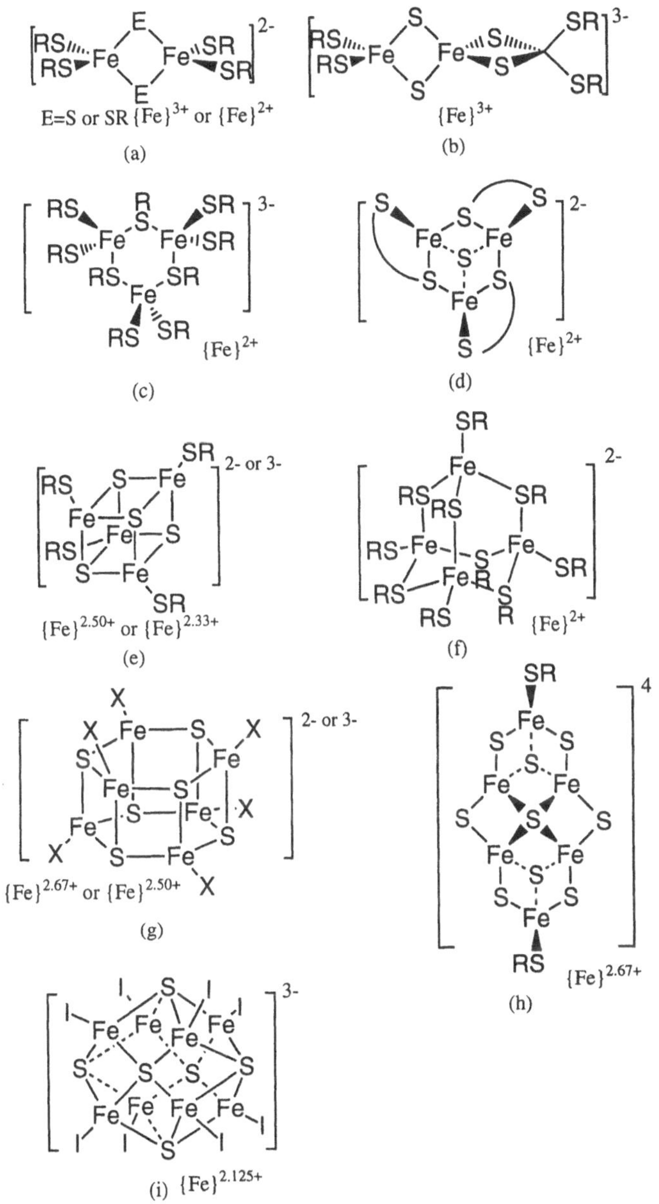

Fig. 19. Various structural types of Fe-S cluster (average oxidation state of Fe is also given)

hydrogenase activities [236]. Through the excellent work over the past decade, numerous polynuclear nickel-chalcogenide cluster complexes have been synthesized and structurally characterized [237–244]. These clusters exhibit diversity in both composition and structure and, consequently, no complete picture of their electronic structures has emerged. As pointed out by Dance and Fisher [8] in their comprehensive review, the diversity of the compositions and geometrical structures known for metal chalcogenide clusters provides real challenges to the fundamental goal of establishing the detailed relationships between observed geometry and bonding theory.

Table 6 shows examples of Ni-S and Ni-Se clusters together with their total valence electron counts and the average oxidation state of Ni for each cluster complex. $Ni_4(\mu_2\text{-Se})(\mu_3\text{-Se})_2(PPh_3)_5$ has a total of 60 valence electrons, following the electron count of tetrahedral carbonyl clusters. $Ni_4(\mu_2\text{-}SC_5H_9NMe)_8$ and $Ni_6(SCH_2CH_2CH_2S)_6$ possess a so-called toroidal structure, and their electronic structures have been studied recently [245] where it was found that the Ni-Ni interactions are not significant. The two distorted cubic structures do not conform to the normal electron count of 120. Their deviations from the normal electron count have been analyzed by Halet and co-workers [219]. The Ni_{12} and Ni_{20} clusters do not follow any electron counting rules developed for other metal clusters. The Ni_{21} cluster has 290 valence electrons, close to the 286 derived

Table 6. Selective examples of some Ni-S or Ni-Se cluster molecules

Cluster Molecule	Electron Count	Average Oxidation State of Ni	Metal Framework	Reference
$Ni_4(\mu_2\text{-}SC_5H_9NMe)_8$	64	+ 2.0	Square	[237]
$Ni_4(\mu_2\text{-Se})(\mu_3\text{-Se})_2(PPh_3)_5$	60	+ 1.5	Tetrahedron	[238]
$[Ni_5(\mu_5\text{-S})(\mu_2\text{-SBu}^t)_5]^-$	70	+ 1.2	Pentagon	[239]
$[Ni_5(\mu_3\text{-Se})_4(PMeEt_2)_6Cl_2$	80	+ 2.0	Vertex-sharing B-triangle	[238]
$Ni_6(SCH_2CH_2CH_2S)_6$	96	+ 2.0	Hexagon	[240]
$Ni_6(\mu_4\text{-Se})_3(\mu_3\text{-Se})_2(PPh_3)_6$	92	+ 1.67	Trigonal Prism	[241]
$Ni_7(\mu_4\text{-Se})_3(\mu_3\text{-Se})_2(PPr^i_3)_6$	102	+ 1.43	Vertex-sharing Bi-tetrahedra	[238]
$[Ni_8(\mu_6\text{-S})(\mu_2\text{-SBu}^t)_9]^-$	114	+ 1.25	Bicapped Trigonal Prism	[237]
$Ni_8(\mu_3\text{-Se})_6(PPr^i_3)_4$	112	+ 1.5	Distorted Cube	[238]
$Ni_8(\mu_3\text{-Se})_6(PPr^i_3)_6$	116	+ 1.5	Distorted Cube	[238]
$[Ni_{12}(\mu_2{:}\eta^2\text{-}S_2)_6(\mu_3\text{-S})_8]^{3-}$	191	+ 2.08	Distorted Cuboctahedron	[242]
$[Ni_{20}(\mu_5\text{-Se})_2(\mu_4\text{-Se})_{10}(\mu_2\text{-SeMe})_{10}]^{2-}$	280	+ 1.6	Capped Pentagonal Antiprism	[243]
$[Ni_{21}(\mu_4\text{-Se})_{14}(PEt_2Ph)_{12}$	290	+ 1.33	[a]	[238]
$[Ni_{34}(\mu_5\text{-Se})_2(\mu_4\text{-Se})_{20}(PPh_3)_{10}]$	448	+ 1.3	[a]	[244]

[a] A central Ni_{13} cuboctahedron is surrounded by two Ni_4 squre units. The two separated squre units are located above two opposite squre faces, respectively, of the cuboctahedron in a prismatic arrangment.

[b] Two sections of **4** are arranged in an approximately anti-prismatic fashion. Between the two sections lies an intervening Ni_4 square.

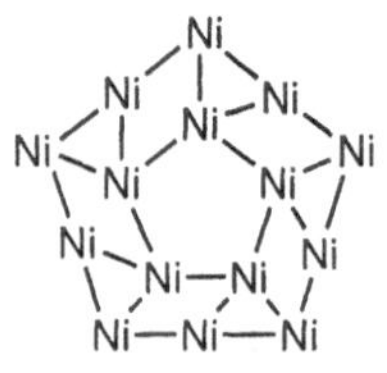

4

from the principle for condensation of clusters from Mingos and Wales [3], while the Ni_{34} cluster follows the topological rules developed by Teo for larger clusters [246]. $[Ni_5(\mu_3\text{-}Se)_4(PMeEt_2)_6Cl_2$ (vertex-sharing bi-triangle), $Ni_7(\mu_4\text{-}Se)_3(m_3\text{-}Se)_2(PPr^i_3)_6$ (vertex-sharing bi-tetrahedra), and $[Ni_8(\mu_6\text{-}S)(\mu_2\text{-}SBu^t)_9]^-$ (bicapped trigonal prism) conform approximately to the principle for condensation of clusters from Mingos and Wales [3]. Clearly the electron count rules developed for other clusters are not so successful for these Ni-S and Ni-Se clusters. A more systematic study of their electronic structures is needed in order to have a better understanding of their structure and bonding relationships.

11 Summary

The discussion presented in this review has demonstrated that metal-metal interactions in early transition metal clusters containing π-donor ligands can be understood in terms of a simple and unified molecular orbital picture. In this molecular orbital picture, one first considers the metal-ligand σ interaction for each metal center and determines which metal orbitals are available for metal-metal bonding. In other words, one should determine which orbitals are used for metal-ligand bonding. Then one can use those available metal orbitals to construct metal-metal orbital interaction diagrams based on orbital overlap and symmetry principles.

In transition metal carbonyl clusters, the so-called "t_{2g}" orbitals are essentially non-bonding due to the $d \rightarrow \pi^*$ backbonding stabilization. These "t_{2g}" sets, however, are the main available orbitals for metal-metal interactions for clusters containing mainly π donating ligands. The active participation of "t_{2g}" orbitals in the metal-metal interactions is a result of the destabilizing effect of π donating ligands. Therefore, one should use a different approach to understand their structure and bonding aspects. It is concluded that d-d interaction plays the most important role in the metal-metal bonding for clusters containing mainly π donating ligands.

The number of metal d electrons in a cluster is determined by the cluster's structural type and the transition metal itself. For earlier transition metal clusters, due to the diffuse nature of their metal d orbitals, the metal-metal antibonding molecular orbitals are normally unoccupied, and the relevant electronic closed-shell requirement are usually met. For late transition metal clusters, the metal-metal interactions are much weaker, and therefore occupation of metal-metal antibonding molecular orbitals is commonly observed. The π-donor ligands in these transition metal clusters not only "activate" the metal "t_{2g}" orbitals for metal-metal bonding, but can also modify the metal-metal orbital levels through their π orbitals; the bridging π-donor ligands also play an important role in binding the metal centers, and greatly affect the metal-metal bond distances for a given formal bond order.

Acknowledgments. We are indebted to the Research Grant Council of Hong Kong and the Hong Kong University of Science & Technology for financial support. ZL would also like to thank Professor D. M. P. Mingos for providing the opportunity of presenting this review article and for his invaluable guidance in the field of cluster chemistry.

References

1. Johnson BFG (ed) (1981) Transition metal clusters. Wiley. New York
2. Lawrence Q (1988) Metal clusters in proteins. ACS symposium series, Washington, DC
3. Mingos DMP, Wales DJ (1990) Introduction to cluster chemistry. Prentice Hall, Englewood Cliffs, New Jersey
4. Shriver DF, Kaesz HD, Adams RD (eds) (1990) The chemistry of metal cluster complexes. VCH, New York
5. Cotton FA, Walton RA (1993) Multiple bonds between metal atoms. Clarendon Press, Oxford
6. Gonzalez-Moraga G (1993) Cluster chemistry: an introduction to the chemistry of transition metal and main group element molecular clusters, Springer, Berlin Heidelberg New York
7. Chisholm MH (ed) (1995) Early transition metal clusters with π-donor ligands. VCH, New York
8. Dance I, Fisher K (1994) Prog Inorg Chem 41:637
9. (a) Williams RE (1971) Inorg Chem 10: 210; (b) Williams RE (1976) Adv Inorg Chem Radiochem 18:67
10. (a) Wade K (1971) J Chem Soc Chem Comm 1971:210; (b) Wade K (1976) Adv Inorg Chem Radiochem 18:1
11. Mingos DMP (1972) Nature Phys Sci 236:99
12. Wade K (1972) Nature Phys Sci 240:71
13. (a) Mingos DMP (1984) Acc Chem Res 17:311; (b) Mingos DMP (1986) Chem Soc Rev 15:31; (c) Mingos DMP, Johnston RL (1987) Structure and Bonding 68:29; (d) Mingos DMP, Slee T, Lin Z (1990) Chem Rev 90:383; (e) Wales DJ, Mingos DMP, Slee T, Lin Z (1990) Acc Chem Res 23:17
14. (a) Stone AJ (1980) Mol Phys 41:1339; (b) Stone AJ (1981) Inorg Chem 20:563; (c) Stone AJ (1984) Polyhedron 3:1229
15. Hoffmann R (1982) Angew Chem Int Ed Engl 21:711
16. Lin Z, Williams ID (1996) Polyhedron 15:3277
17. McPartlin M, Eady CR, Johnson BFG, Lewis J (1976) J Chem Soc Chem Comm 1976:883
18. (a) Bertrand JA, Cotton FA, Dollase WA (1963) J Am Chem Soc 85:1349; (b) Cotton FA, Haas TE (1964) Inorg Chem 3:10
19. Cotton FA, Wilkinson G (1988) Advanced inorganic chemistry, 5th edn, Wiley, New York, p1085
20. Kraatz HB, Boorman PM (1995) Coord Chem Rev 143:35
21. Cotton FA, Daniels L, Davison A, Orvig C (1981) Inorg Chem 20:3051
22. Koz'min PA, Larina TB, Surazhskaya MD (1981) Koord Khim 7:1719
23. Koz'min PA, Larina TB, Surazhskaya MD (1982) Koord Khim 8:851
24. Burns CJ, Burrell AK, Cotton FA, Haefner SC, Sattelberger AP (1994) Inorg Chem 33:2257
25. Cotton FA, Dunbar KR, Falvello LR, Tomas M, Walton RA (1983) J Am Chem Soc 105:4950
26. Huang HW, Martin DS (1985) Inorg Chem 24:96
27. Cotton FA, Matusz M (1987) Inorg Chem 26:3468
28. Cotton FA, Jennings JG, Price AC, Vidyasagar K (1990) Inorg Chem 29:4138
29. Bino A, Cotton FA, Felthouse TR (1979) Inorg Chem 18:2599
30. Lindsay AJ, Tooze RP, Motevalli M, Hursthouse MB, Wilkinson G (1984) J Chem Soc, Chem Comm 1984:1383
31. Agaskar PA, Cotton FA, Dunbar KR, Falvello LR, Tetrick SM, Walton RA (1986) J Am Chem Soc 108:4850
32. Cotton FA, Dunne TG, Wood JS (1964) Inorg Chem 3:1495
33. Cotton FA, Felthouse TR (1980) Inorg Chem 19:323
34. Bellitto C, Dessy G, Fares V (1985) Inorg Chem 24:2815
35. Kullberg ML, Lemke FR, Powell DR, Kubiak CP (1985) Inorg Chem 24:3589

36. Appleton TG, Byriel KA, Hall JR, Kennard CHL, Mathieson MT (1992) J Am Chem Soc 114:7305
37. Bellitto C, Flamini A, Piovesana O, Zanazzi PF (1980) Inorg Chem 19:3632
38. Manojlovic-Muir L, Muir KW, Solomun T (1979) Acta Crystallogr Sect B35:1237
39. Sobota P, Ejfler J, Szafert S, Szczegot K, Sawaka-Dobrowlska W (1993) J Chem Soc Dalton Trans 1993:2353
40. Wengrovius JH, Schrock RR, Day CS (1981) Inorg Chem 20:1844
41. Cotton FA, Shang M, Wojtczak WA (1991) Inorg Chem 30:3670
42. Floriani C, Gambarotta S, Chiesi-Villa A, Guastini C (1987) J Chem Soc Dalton Trans 1987:2099
43. Cotton FA, Lu J, Ren T (1994) Inorg Chim Acta 215:47
44. Nieman J, Teuben JH (1986) Organometallics 5:1149
45. Visciglio VM, Fanwick PE, Rothwell IP (1994) Acta Cryst C50:900
46. Drew MGB, Rice DA, Williams DM (1985) J Chem Soc Dalton Trans 1985:417
47. Cotton FA, Roth WJ (1983) Inorg Chem 22:3654
48. Preuss F, Lambing G, Muller-Becker S (1994) Z Anorg Allg Chem 620:1812
49. Babaian-Kibala E, Cotton FA, Kibala PA (1990) Inorg Chem 29:4002
50. Cotton FA, Falvello LR, Najjar RC (1983) Inorg Chem 22:375
51. Cotton FA, Eglin JL, Luck RL, Son K (1990) Inorg Chem 29:1802
52. DuBois MR, DuBois DL, Derveer MC, Haltiwanger RC (1981) Inorg Chem 20: 3064
53. Chisholm MH, Kirkpatrick CC, Huffman JC (1981) Inorg Chem 20:871
54. Hey E, Weller F, Dehnicke K (1984) Z Anorg Allg Chem 508:86
55. Cotton FA, Powell GL (1984) J Am Chem Soc 106:3371
56. Rau MS, Kretz CM, Geoffroy GL, Rheingold AL (1993) Organometallics 12:3447
57. Anderson LB, Cotton FA, DeMarco D, Fang A, Ilsley WH, Kolthammer BWS, Walton RA (1981) J Am Chem Soc 103:5078
58. Dilworth JR, Richards RL, Dahlstrom P, Hutchinson J, Kumar S, Zubieta J (1983) J Chem Soc Dalton Trans 1983:1489
59. Chacon ST, Chisholm MH, Streib WE, van der Sluys W (1989) Inorg Chem 28:5
60. Willett RD (1979) Acta Crystallogr Sect B35: 178
61. Bryan JC, Wheeler DR, Clark DL, Huffman JC, Sattelberger AP (1991) J Am Chem Soc 113:3184
62. Barder TJ, Cotton FA, Lewis D, Schwotzer W, Tetrick SM, Walton RA (1984) J Am Chem Soc 106:2882
63. Herrmann WA, Fischer RA, Felixberger JK, Paciello RA, Kiprof P, Herdtweck E (1988) Z Naturforsch Teil B43:1391
64. Bentrup U, Massa W (1991) Z Naturforsch Teil B46:395
65. Cotton FA, Matusz M, Torralba RC (1989) Inorg Chem 28:1516
66. McCormick FB, Gleason WB (1988) Acta Cryst C44:603
67. Krebs B, Henkel G, Dartmann M, Preetz W, Bruns M (1984) Z Naturforsch Teil B39:843
68. Gross CL, Wilson SR, Girolami GS (1994) J Am Chem Soc 116:10|294
69. Watkins SF, Fronczek FR (1982) Acta Crystallogr B38:270
70. Mueting AM, Boyle P, Pignolet LH (1984) Inorg Chem 23:44
71. Muir JA, Muir MM, Rivera AJ (1974) Acta Crystallogr B30:2062
72. Churchill MR, Julis SA (1977) Inorg Chem 16:1488
73. Landee CP, Willett RD (1981) Inorg Chem 20:2521
74. Solari E, Floriani C, Chiesi-Villa A, Guastini C (1989) J Chem Soc Chem Comm 1989:1747
75. Cotton FA, Wojtczak WA (1994) Inorg Chim Acta 216:9
76. Scherfise KD, Willing W, Muller U, Dehnicke K (1986) Z Anorg Allg Chem 534:85
77. Rambo JR, Bartley SL, Streib WE, Christou G (1994) J Chem Soc Dalton Trans 1994:1813
78. Cotton FA, Shang M (1993) Inorg Chem 32:969
79. Canich JAM, Cotton FA (1989) Inorg Chim Acta 159:163
80. Cotton FA, Najjar RC (1981) Inorg Chem 20:2716
81. Kiriazis L, Mattes R (1991) Z Anorg Allg Chem 593:90

82. Cotton FA, Poli R (1987) Inorg Chem 26:3310
83. Boorman PM, Codding PW, Kerr KA, Moynihan KJ, Patel VD (1982) Can J Chem 60:1333
84. Ball JM, Boorman PM, Moynihan KJ, Patel VD, Richardson JF, Collison D, Mabbs FE (1983) J Chem Soc Dalton Trans 1983:2479
85. Boorman PM, Patel VD, Kerr KA, Codding PW, Roey P (1980) Inorg Chem 19:3508
86. Kiriazis L, Mattes R (1991) Z Anorg Allg Chem 593:90
87. Cotton FA, Torralba RC (1991) Inorg Chem 30:2196
88. Cotton FA, Kang SJ (1993) Inorg Chem 32:2336
89. Duncan JAS, Hedden D, Roundhill DM, Stephenson TA, Walkinshaw MD (1982) Angew Chem Int Ed Engl 21:452
90. Nutton A, Bailey PM, Maitlis PM (1981) J Chem Soc Dalton Trans 1981:1997
91. Janas Z, Lis T, Sobota P (1992) Polyhedron 11:3019
92. Calderazzo F, Pampaloni G, Rocchi L, Strahle J, Wurst K (1991) J Organomet Chem 413: 91
93. Ting C, Baenziger NC, Messerle L (1988) J Chem Soc Chem Comm 1988:1133
94. Brunner H, Meier W, Wachter J, Weber P, Ziegler ML, Enemark JH, Young CG (1986) J Organomet Chem 309:313
95. Brunner H, Pfauntsch J, Wachter J, Nuber B, Ziegler ML (1989) J Organomet Chem 359: 179
96. Brunner H, Meier W, Wachter J, Guggolz E, Zahn T, Ziegler ML (1982) Organometallics 1: 1107
97. Green MLH, Izquierdo A, Martin-Polo JJ, Mtetwa VSB, Prout K (1983) J Chem Soc Chem Comm 1983:538
98. Miller WK, Haltiwanger RC, VanDerveer MC, DuBois MR (1983) Inorg Chem 22:2973
99. Harlan CJ, Jones RA, Koschmieder SU, Nunn CM (1990) Polyhedron 9:669
100. Cotton FA, Roth WJ (1984) Inorg Chem 23:945
101. Cotton FA, Diebold MP, Roth WJ (1985) Polyhedron 4: 103
102. Fenske D, Czeska B, Schumacher C, Schmidt RE, Dehnicke K (1985) Z Anorg Allg Chem 520:7
103. Chetcuti MJ, Chisholm MH, Folting K, Huffman JC, Janos J (1982) J Am Chem Soc 104: 4684
104. Chisholm MH, Corning JF, Huffman JC (1984) Inorg Chem 23:754
105. Gilbert TM, Landes AM, Rogers RD (1992) Inorg Chem 31:3438
106. Chisholm MH, Cotton FA, Extine MW, Millar M, Stults BR (1977) Inorg Chem 16:320
107. Bursten BE, Cotton FA, Green JC, Seddon EA, Stanley GG (1980) J Am Chem Soc 102:4579
108. (a) Müller A (1986) Polyhedron 5:23 and references cited therein; (b) Chakrabarty PK, Ghost I, Bhattacharyya R, Mukherjee AK, Mukherjee M, Helliwell M (1996) Polyhedron 15:1443 and references cited therein
109 Saito T (1995) In: Chisholm MH (ed) Early transition metal clusters with π-donor ligands. VCH, New York, p 63
110. Jiang Y, Tang A, Hoffmann R, Huang J, Lu J (1985) Organometallics 4:27
111. Cotton FA, MagueJT (1964) Inorg Chem 3:1094
112. Irmler M, Meyer G (1991) Z Anorg Allg Chem 604:17
113. Andres R, Galakhov MV, Martin A, Mena M, Santamaria C (1994) Organometallics 13: 2159
114. Cotton FA, Diebold MP, Llusar R, Roth WJ (1986) J Chem Soc Chem Comm 1986:1276
115. Cotton FA, Kibala PA, Roth WJ (1988) J Am Chem Soc 110:298
116. Cotton FA, Diebold MP, Roth WJ (1987) J Am Chem Soc 109:2833
117. Richeson DS, Hsu SW, Fredd NH, Duyne GV, Theopold KH (1986) J Am Chem Soc 108: 8273
118. Richens DT, Helm L, Pittet PA, Merbach AE, Nicolo F, Chapuis G (1989) Inorg Chem 28: 1394
119. Cotton FA, Llusar R (1987) Polyhedron 6:1741
120. Neumann HP, Ziegler ML (1988) J Chem Soc Chem Comm 1988:498
121. Cramer RE, Yamada K, Kawaguchi H, Tatsumi K (1996) Inorg Chem 35:1743

122. Cotton FA, Shang M, Sun ZS (1991) J Am Chem Soc 113:3007
123. Cotton FA, Shang M, Sun ZS (1991) J Am Chem Soc 113:6917
124. Shibahara T, Yamasaki M, Sakane G, Minami K, Yabuki T, Ichimura A (1992) Inorg Chem 31:640
125. Cotton FA, Shang M, Sun ZS (1992) J Cluster Sci 3:123
126. Müller U, Krug V (1988) Angew Chem Int Ed Engl 27:293
127. Dean NS, Folting K, Lobkovsky E, Christou G (1993) Angew Chem Int Ed Engl 32:594
128. Klingelhofer P, Muller U, Friebel C, Pebler J (1986) Z Anorg Allg Chem 543:22
129. Fedin VP, Muller A, Filipek K, Rohlfing R, Bogge H, Virovets AV, Dziegielewski JO (1994) Inorg Chim Acta 223:5
130. Fedin VP, Imoto H, Saito T (1995) J Chem Soc Chem Comm 1995:1559
131. Fedin VP, Sokolov MN, Geras'ko OA, Kolesov BA, Fedorov VY, Mironov AV, Yufit DS, Slovohotov YL, Struchkov YT (1990) Inorg Chim Acta 175:217
132. Mizobe Y, Hashizume K, Murai T, Hidai M (1994) J Chem Soc Chem Comm 1994:1051
133. Lockemeyer JR, Rauchfuss TB, Rheingold AL (1989) J Am Chem Soc 111:5733
134. Casey CP, Widenhoefer RA, Hallenbeck SL, Hayashi RK, Powell DR, Smith GW (1994) Organometallics 13:1521
135. Pulliam CR, Thoden JB, Stacy AM, Spencer B, Englert MH, Dahl LF (1991) J Am Chem Soc 113:7398
136. Kamijo N, Watanabe T (1979) Acta Crystallogr B35:2537
137. de Miguel AV, Isobe K, Bailey PM, Meanwell NJ, Maitlis PM (1982) Organometallics 1: 1604
138. Nishioka T, Isobe K (1994) Chem Lett 1994:1661
139. Venturelli A, Rauchfuss TB (1994) J Am Chem Soc 116:4824
140. Edema JJH, Duchateau R, Gambarotta S, Bensimon C (1991) Inorg Chem 30:3585
141. Hughes DL, Larkworthy LF, Leigh GJ, McGarry CJ, Sanders JR, Smith GW, de Souza JS (1994) J Chem Soc Chem Comm 1994:2137
142. Arif AM, Hafner JG, Jones RA, Albright TA, Kang SK (1986) J Am Chem Soc 108:1701
143. Chisholm MH, Folting K, Huffman JC, Kirkpatrick CC (1984) Inorg Chem 23:1021
144. Tsuge K, Yajima S, Imoto H, Saito T (1992) J Am Chem Soc 114:7910
145. Cotton FA, Duraj SA, Roth WJ (1984) J Am Chem Soc 106:3527
146. Ardon M, Bino A, Cotton FA, Dori Z, Kaftory M, Reisner G (1982) Inorg Chem 21:1912
147. Bino A, Hesse KF, Kuppers H (1980) Acta Crystallogr B36:723
148. Chisholm MH (1995) In: Chisholm MH (ed) Early transition metal clusters with π-donor ligands. VCH, New York, p165
149. Seela JL, Huffman JC, Christou G (1987) J Chem Soc Chem Comm 1987:1258
150. Babaian-Kibala E, Cotton FA, Kibala PA (1990) Polyhedron 9:1689
151. Chisholm MH, Clark DL, Errington RJ, Folting K, Huffman JC (1988) Inorg Chem 27:2071
152. Cotton FA, Powell GL (1983) Inorg Chem 22:871
153. Aufdembrink BA, McCarley RE (1986) J Am Chem Soc 108:2474
154. Jiang F, Lei X, Huang Z, Hong M, Kang B, Wu D, Liu H (1990) J Chem Soc Chem Comm 1990:1655
155. Kriege M, Henkel G (1987) Z Naturforsch Teil B, 42:1121
156. Fenske D, Hollnagel A, Merzweiler K (1988) Angew Chem Int Ed Engl 27:965
157. Mimoun H, Charpentier R, Mitschler A, Fischer J, Weiss R (1980) J Am Chem Soc 102: 1047
158. Carrondo MA, Skapski AC (1978) Acta Crystallogr Sect B 34:3576
159. Wheeler RA, Hoffmann R (1986) J Am Chem Soc 108:6605
160. Babaian-Kibala E, Cotton FA (1991) Acta Cryst C 47:1716
161. Tsuge K, Imoto H, Saito T (1992) Inorg Chem 31:4715
162. Chisholm MH, J Errington RJ, Folting K, Huffman JC (1982) J Am Chem Soc 104:2025
163. Tsuge K, Imoto H, Saito T (1993) Inorg Chem 32:1321
164. Chisholm MH, Clark DL, Folting K, Huffman JC (1986) Angew Chem Int Ed Engl 25: 1014
165. Budzichpwski TA, Chisholm MH, Folting K, Steib WE, Scheer M (1996) Inorg Chem 35: 3659

166. Chisholm MH, Huffman JC, Leonelli J (1981) J Chem Soc Chem Comm 1981:270
167. Chisholm MH, Folting K, Huffman JC, Leonelli J, Marchant NS, Smith CA, Taylor LC (1985) J Am Chem Soc 107:3722
168. Toan T, Teo BK, Ferguson JA, Meyer TJ, Dahl LF (1977) J Am Chem Soc 99:408
169. Darkwa J, Lockemeyer JR, Boyd PD, Rauchfuss TB, Rheingold AL (1988) J Am Chem Soc 110:141
170. Pasynskii AA, Eremenko IL, Rakitin YV, Novotortsev VM, Ellert OG, Kalinniko VT, Shklover VE, Struchkov YT, Lindeman SV, Kurbanov TK, Gasanov GS (1983) J Organomet Chem 248:309
171. Bandy JA, Davies CE, Green JC, Green MLH, Prout K, Rodgers DPS (1983) J Chem Soc Chem Comm 1983:1395
172. Müller A, Jostes R, Eltzner W, Nie CS, Diemann E, Bögge H, Zimmermann M, Dartmann M, Reinsch-Vogell U, Che S, Cyvin SJ, Cyvin BN (1985) Inorg Chem 24:2872
173. Fedorov VE, Mironov YV, Fedin VP, Imoto H, Saito T (1996) Acta Cryst C52:1065
174. Cotton FA, Lu J, Shang M, Wojtczak WA (1994) J Am Chem Soc 116:4364
175. Bottomley F, Drummond DF, Egharevba GO, White PS (1985) Organometallics 5:1620
176. Bottomley F, Day RW (1996) Organometallics 15:809
177. Fenske D, Grissinger A (1990) Z Naturforsch 45B: 1309
178. (a) Bottomley F, White PS (1981) J Chem Soc Chem Comm 1981:28; (b) Bottomley F, Paez DE, White PS (1982) J Am Chem Soc 104:5651
179. Bolinger CM, Darkwa J, Gammie G, Gammon SD, Lyding JW, Rauchfuss TB, Wilson SR (1986) Organometallics 5:2386
180. Ziebarth RP, Corbett JD (1989) Accts Chem Res 22:256
181. Kohler J, Svensson G, Simon A (1992) Angew Chem Int Ed Engl 31:1437
182. Lee SC, Holm RH (1990) Angew Chem Int Ed Engl 29:840
183. (a) Hughbanks T, Hoffmann R (1983) J Am Chem Soc 105:1150; (b) Hughbanks T, Rosenthal G, Corbett JD (1986) J Am Chem Soc 108:8289; (c) Bond MR, Hughbanks T (1992) Inorg Chem 31:5015
184. (a) Johnston RL, Mingos DMP (1986) Inorg Chem 25:1661; (b) Mingos DMP, Lin Z (1989) Z Phys D Atoms, Molecules and Clusters 12:53
185. Cotton FA, Hughbanks T, Runyan CE, Wojtczak WA (1995) In: Chisholm MH (ed) Early transition metal clusters with π-donor ligands. VCH, New York, p1
186. Saito T, Yamamoto N, Yamagata T, Imoto H (1988) J Am Chem Soc 110:1646
187. Saito T, Yamamoto N, Nagase T, Tsuboi T, Kobayashi K, Yamagata T, Imoto H, Unoura K (1990) Inorg Chem 29:764
188. Ouahab L, Batail P, Perrin C, Garrigou-Lagrange C (1986) Mater Res Bull 21:1223
189. Boeschen S, Keller HL (1990) Z Kristallogr 200:305
190. Saito T, Yoshikawa A, Yamagata T, Imoto H,Unoura K (1989) Inorg Chem 28:3588
191. Saito T, Manabe H, Yamagata T, Imoto H (1987) Inorg Chem 26:1362
192. Gabriel JC, Boubekeur K, Batail P (1993) Inorg Chem 32:2894
193. Yaghi OM, Scott MJ, Holm RH (1992) Inorg Chem 31:4778
194. Mironov YU, Virovets AV, Fedorov VE, Podberezskaya NV, Shishkin OV, Struchkov YT (1995) Polyhedron 14:3171
195. Agresti A, Bacci M, Cecconi F, Ghilardi CA, Midollini S (1985) Inorg Chem 24:689
196. Fenske D, Hachgenei J, Ohmer J (1985) Angew Chem Int Ed Engl 24:706
197. Diana E, Gervasio G, Rossetti R, Valdemarin F, Bor G, Stanghellini PL (1991) Inorg Chem 30:294
198. Steigerwald ML, Siegrist T, Stuczynski SM (1991) Inorg Chem 30:2256
199. Fenske D, Grissinger A, Loos M, Magull J (1991) Z Anorg Allg Chem 598/599:121
200. Koknat FW, McCarley R E (1972) Inorg Chem 11:812
201. Koknat FW, McCarley RE (1974) Inorg Chem 13:295
202. Ueno F, Simon A (1985) Acta Crystallogr C 41:308
203. Beck U, Borrmann H, Simon A (1994) Acta Cryst C 50:695
204. Brnicevic N, Nothig-Hus D, Kojic-Prodic B, Ruzic-Toros Z, Danilovic Z, McCarley RE (1992) Inorg Chem 31:3924

205. Basson SS, Leipoldt G (1982) Trans Met Chem 7:207
206. Dudis DD, Corbett JD, J Hwu S (1986) Inorg Chem 25:3434
207. Hinz DJ, Meyer G (1994) J Chem Soc Chem Comm 1994:125
208. Zhang J, Ziebarth RP, Corbett JD (1992) Inorg Chem 31:614
209. Zhang J, Corbett JD (1993) Inorg Chem 32:1566
210. Qi RY, Corbett JD (1995) Inorg Chem 34:1646
211. Hughbanks T, Corbett JD (1988) Inorg Chem 27:2022
212. Bencini A, Ghilardi GA, Midollini S, Orlandini A, Russo U, Uytterhoeven MG, Zanchini C (1995) J Chem Soc Dalton Trans 1995:963
213. (a) Spitzin VI, Kuzina AF, Oblova AA, Krynchkov SV (1985) Russ Chem Rev 54:373; (b) Koz'min PA, Surazhskaya MD, Larina TB (1985) Koord Khim 11:1559
214. (a) Spitzin VI, Krynchkov SV, Grigoriev MS, Kuzina AF (1988) Z anorg allg Chem 563: 136; (b) Krynchkov SV, Kuzina AF, Spitzin VI (1988) Z anorg allg Chem 563:153
215. Koz'min PA, Kotel'nikova AS, Larina TB, Mekhtiev MM, Surazhskaya MD, Bagiro v SA, Osmanov NS (1987) Dokl Akad Nauk SSSR 295:647
See question here
216. Koz'min PA, Kotelnikova AS, Surazhskaya MD, Osmanov, Larina TB, Abbasova TA, Mekhtiev MM (1989) Koord Khim 15:1216
217. Womelsdorf H, Meyer HJ (1994) Angew Chem Int Ed Engl 33:1943
218. (a) Kerihuel G, Gougeon P (1995) Acta Cryst C51: 789; C51: 1475; (b) Gougeon P, McCarley RE (1991) Acta Cryst C47:241
219. (a) Furet E, Beuze AL, Halet JF, Saillard J Y (1994) J Am Chem Soc 116:274; (b) Furet E, Beuze AL, Halet JF, Saillard J Y (1995) J Am Chem Soc 117:4936
220. Schmek GL, Chen SC, McCarley RE (1995) Inorg Chem 34:6130
221. (a) Kim J, Rees DC (1992) Nature 360:553; Science 257:1677; (b) Chan MK, Kim J, Rees DC (1993) Science 260:792
222. (a) Holm RH (1992) Adv Inorg Chem 38:1; (b) Coucouvanis D (1991) Acc Chem Res 24:1
223. Coucouvanis D (1981) Acc Chem Res 14:201
224. (a) Müller A, Hellmann W, Bögge H, Jostes R, Römer M, Schimanski U (1982) Angew hem Int Ed Engl 21: 860; (b) Müller A, Hellmann W, Römer C, Römer M, Bögge H, Jostes R Schimanski U (1984) Inorg Chim Acta 83:L75; (c) Müller A, Sarkar S (1977) Angew Chem Int Ed Engl 16:705
225. Chen CN et al. (1996) unpublished result
226. (a) Deng H, Hoffmann R (1993) Angew Chem Int Ed Engl 32:1062; (b) Stavrev KK, Zerner MC (1996) Chem Eur J 2:83; (c) Liu C, Lin Z, Lu J (1988) J Mol Struct (Theochem) 180:189; (d) Liu C, Hua J, Chen Z, Lin Z, Lu J (1986) Int J Quantum Chem 29:701; (e) Bowmaker GA, Boyd PD, Sorrenson RJ, Reed CA, McDonald JW (1987) Inor Chem 26:3; (f) Noodleman L (1991) Inorg Chem 30:246, 256; (g) Bertini I, Ciurli S, Luchinat C (1995) Structure and Bonding 83:1
227. Tieckelmann RH, Averill BA (1980) Inorg Chim Acta 46:L35
228. Simopoulos A, Papaefthymiou V, Kostikas A, Petrousleas V, Coucouvanis D, Simhon ED, Stremple P (1981) Chem Phys Lett 81:261
229. Hagen KS, Watson AD, Holm RH (1983) J Am Chem Soc 105:3905
230. Hagen KS, Christou G, Holm RH (1983) Inorg Chem 22:309
231. Berg JM, Holm RH (1982) In: Spiro TG (ed) Iron-sulfur proteins. Wiley, New York, p1
232. Hagen KS, Holm RH (1984) Inorg Chem 23:418
233. Kanatzidis MG, Salifoglou A, Coucouvanis D (1986) Inorg Chem 25:2460
234. Strasdeit H, Krebs B, Henkel G (1984) Inorg Chem 23:1816
235. Saak W, Pohl S (1991) Angew Chem Int Ed Engl 30:881
236. Lancaster JR (Ed) (1988) The bioinorganic chemistry of nickel. VCH, Weinheim
237. Krüger T, Krebs B, Henkel G (1989) Angew Chem Int Ed Engl 28:61
238. Fenske D, Krautscheid H, Müller M (1992) Angew Chem Int Ed Engl 31:321
239. Müller A, Henkel G (1996) J Chem Soc Chem Comm 1996:1005
240. Sletten J, Kovacs JA (1994) Acta Chemica Scand 48:929

241. Fenske D, Ohmer J (1987) Angew Chem Int Ed Engl 26:148
242. Cahill CL, Tan K, Novoseller R, Parise JB (1996) J Chem Soc Chem Comm 1996:1677
243. Fenske D, Fischer A (1995) Angew Chem Int Ed Engl 34:307
244. Fenske D, Ohmer J, Hachgenei J (1985) Angew Chem Int Ed Engl 24:993
245. Alemany P, Hoffmann R (1993) J Am Chem Soc 115:8290
246. Teo BK (1983) J Chem Soc Chem Comm 1983:1362

Electron Count Versus Structural Arrangement in Clusters Based on a Cubic Transition Metal Core with Bridging Main Group Elements

Jean-François Halet and Jean-Yves Saillard

Laboratoire de Chimie du Solide et Inorganique Moléculaire, URA CNRS 1495, Université de Rennes 1, 35042 Rennes Cedex, FRANCE. Fax: + (0)299635704; *E-mail: halet@univ-rennes1.fr* and *saillard@univ-rennes1.fr*

The bonding of diverse molecular clusters based on a cubic transition metal core with bridging main group elements is reviewed. It is shown that the cluster topology of these compounds is dependent on several parameters which can be calibrated. A large range of electron counts is generally observed without alteration of the cubic architecture. Examination of their electronic structures indicates that these species are at the borderline between molecular and solid states, characterized either by a closed-shell electronic configuration secured by a significant HOMO-LUMO gap, or by a quasi-continuous energy level structure often associated with open-shell electronic configurations.

Keywords: Electron counting rules; electronic structure; cubic transition cluster

1 Introduction

Since Cotton first coined the word "metal clusters" in the mid-sixties to describe chemical systems containing metal-metal bonds [1], many such compounds have been synthesized and characterized. These metal-metal bound systems are generally classified in two broad families which often differ from each other in many ways: those having π-acceptor ligands surrounding the metal core such as carbonyls, phosphines, or organic π-systems, and those possessing π-donor ligands such as phosphido groups, chalcogenides, or halides. The octahedral species $[Os_6(CO)_{18}]^{2-}$ and $[Mo_6(\mu_3\text{-Cl})_8Cl_6]^{2-}$ provide examples belonging to the former and the latter, respectively [2, 3]. Despite the vast structural variety encountered in these compounds, general patterns have emerged rapidly which have allowed the development of various localized or delocalized descriptions of their bonding. Today, the set of electron counting rules known as the Polyhedral Skeletal Electron Pair (PSEP) theory, first introduced 25 years ago [4], is probably the most popular bonding pattern used. It has proved to be particularly powerful for rationalizing or predicting the geometrical features of many metal clusters having closed-shell electron configurations, particularly with π-acid systems. However, it has been less successfully applied to clusters containing π-donor ligands, whose electronic structures result in small HOMO-LUMO gaps and are associated with open-shell electron configurations. In order to circumvent the limitations of the PSEP theory in rationalizing this kind of cluster, quantum-mechanical calculations are particularly useful for understanding their structural arrangement and their properties, and eventually for revising the electron-counting rules.

Hitherto, the majority of the theoretical work devoted to this kind of compound has been within the extended Hückel (EH) formalism [5]. This one-electron non-self-consistent method has proved to be particularly effective for giving trends concerning the bonding and the reactivity of molecular compounds. This is due to the possibility of rapidly providing sets of results which can be directly interpretable in terms of perturbation theory arguments which emphasize the interactions between frontier orbitals of molecular fragments. However, its limitations such as the difficulty in optimizing bond distances or the treatment of open-shell configurations have now led theoretical chemists to complement EH studies with more quantitative approaches. Among them, the methods based on density functional theory (DFT) [6], which models electron correlation via functionals of the electron density, have gained steadily in popularity in the last few years. This stems in large measure from its computational expedience compared to post-Hartree-Fock methods, which allows chemists to study the electronic structure of large-size transition metal compounds with reasonable accuracy.

The octahedral M_6 structural arrangement is by far the most documented in π-donor transition-metal cluster chemistry. For details concerning the numerous theoretical studies of compounds based on such an arrangement, the reader should refer to the vast amount of information available in many papers [7]. The cubic M_8 architecture represents an alternative arrangement which is far less commonly encountered up to now. However, with the impetus provided by the research groups of Dahl, Holm in US, Fenske and Pohl in Germany, the number of compounds containing a cubic M_8 core has increased rapidly. They now represent an important family within inorganic transition-metal cluster chemistry [8]. This review deals with the bonding and the rationalization of these species.

2
Noncentered Hexacapped $M_8(\mu_4\text{-}E)_6L_n$ Species

$Ni_8(\mu_4\text{-}PPh)_6(CO)_8$ (1), prepared by Lower and Dahl 20 years ago [9] is a typical representative example of the noncentered cubic inorganic transition-metal clusters of formula $M_8(\mu_4\text{-}E)_6L_n$ (M = transition metal, E = main group element, L = two-electron terminal ligand (CO, PR_3, Cl^-...) and $n \leq 8$). In this compound, the metal atoms form a regular cube, the faces of which are capped by π-donor phosphido groups. Every metal atom bears additionally a terminal carbonyl ligand, leading to a tetrahedral ligand environment. The Ni-Ni separations are rather short, 2.65 Å, indicating that metal-metal bonding is present in 1. Such a

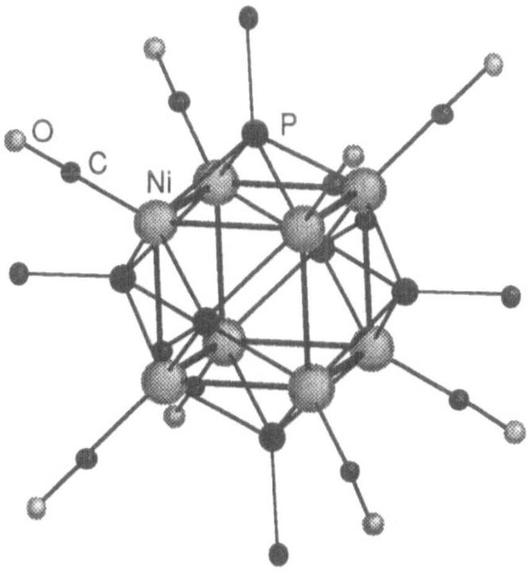

$Ni_8(\mu_4\text{-}PPh)_6(CO)_8$

1

complex belongs to the family of capped three-connected polyhedral clusters which, as stated by Johnston and Mingos, are generally characterized by 3n skeletal electrons (n represents the number of metallic centers of the cluster cage) or 15n metallic valence electrons (MVE) [10]. The count of 24 skeletal electrons or 120 MVEs for **1** [10 (Ni) × 8 + 2 (CO) × 8 + 4 (PPh) × 6 = 15 × 8 = 120) is then in agreement with these electron-counting rules.

For several years this compound was regarded as a laboratory curiosity. Today there are about 20 noncentered hexacapped cubic compounds which have been structurally characterized with Ni, Co, and Fe. They are given in Table 1. They can be partitioned into two groups. The first one is made of compounds such as **1** bearing a terminal ligand on each metal center (n=8). Compounds presenting an incomplete shell of terminal ligands (n < 8) such as $Ni_8(\mu_4\text{-}Se)_6(P\textit{i}Pr_3)_4$ (**2**) [14] constitute the second group. There are a few examples which, in common with **1**, have the expected electron count of 120, but

Table 1. Molecular cubic $M_8(\mu_4\text{-}E)_6L_n$ clusters characterized by X-ray diffraction

Compound	$d_{M\text{-}M}$ [a]	MVE [b]	Ref.
a) $M_8(\mu_4\text{-}E)_6L_8$			
$Ni_8(\mu_4\text{-}PPh)_6(CO)_8$	2.65	120	9
$Ni_8(\mu_4\text{-}PPh)_6(PPh_3)_4(CO)_4$	2.67	120	11
$Ni_8(\mu_4\text{-}PPh)_6(CO)_4(AsPh_3)_4$		120	12
$Ni_8(\mu_4\text{-}S)_6(PPh_3)_8$	2.70	120	13
$Ni_8(\mu_4\text{-}Se)_6(P\textit{n}Bu_3)_8$	2.70	120	14
$Ni_8(\mu_4\text{-}S)_6(PPh_3)_6Cl_2$	2.68	118	15
$Ni_8(\mu_4\text{-}PPh)_6(PPh_3)_4Cl_4$	2.61	116	11
$Ni_8(\mu_4\text{-}PPh)_6(PPh_3)_4Br_4$	2.61	116	11
$[Ni_8(\mu_4\text{-}PPh)_6(PPh_3)_4Cl_4]^+$	2.58	115	16
$Ni_5Fe_3(\mu_4\text{-}S)_6I_8{}^{4-}$	2.68	110	17
$[Co_8(\mu_4\text{-}S)_6(SPh)_8]^{5-}$	2.67	109	18
$[Co_8(\mu_4\text{-}S)_6(SPh)_8]^{4-}$	2.66	108	18
$Fe_4Ni_4(\mu_4\text{-}S)_6I_4(PMePh_2)_4$	2.65	108	19
$[Fe_6Ni_2(\mu_4\text{-}S)_6I_6(PMePh_2)_2]^{2-}$	2.72/2.65	104	19
$[Fe_8(\mu_4\text{-}S)_6I_8]^{4-}$	2.72	100	20
$[Fe_8(\mu_4\text{-}S)_6I_8]^{3-}$	2.72	99	21
b) $M_8(\mu_4\text{-}E)_6L_n$ (n < 8)			
$Ni_8(\mu_4\text{-}Se)_6(PEt_2Ph)_6$		116	14
$Ni_8(\mu_4\text{-}PPh)_6(PPh_3)_4Ni$	2.54	114	15
$Ni_8(\mu_4\text{-}PPh)_6(PPh_3)_4Hg$	2.53	114	13
$Ni_8(\mu_4\text{-}Se)_6(P\textit{i}Pr_3)_4$		112	14
$Ni_8(\mu_4\text{-}PPh)_6(PPh_3)_4$	2.53	112	22
$Ni_8(\mu_4\text{-}P\textit{t}Bu)_6(PPh_3)_2$	2.50	110	12
c) $M_8(\mu_4\text{-}E)_6L_n$ (n > 8)			
$Rh_8(\mu_4\text{-}P(C_5H_9))_6(CO)_9$		114	23

[a] Averaged metal-metal distance (Å).
[b] Total valence metallic electron count.

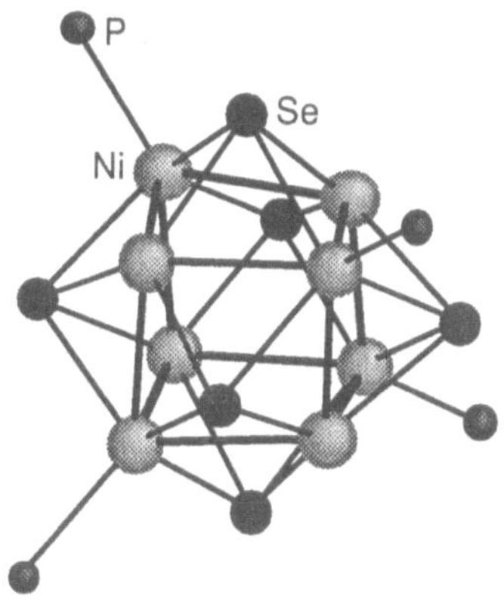

$Ni_8(\mu_4\text{-}Se)_6(PiPr_3)_4$

2

there are many possessing less. Indeed, a wide range from 99 to 120 MVEs is observed so far for the same relatively regular cubic architectural unit. Just a slight shrinking of the metallic core is noticed when the number of MVEs decreases. This is somewhat surprising at first sight. In molecular transition metal chemistry we are used to associate bond breaking or bond making with addition or removal of electrons. How can this be rationalized?

Several theoretical studies have been reported on the compounds given in Table 1. Some years ago, Burdett and Miller studied the bonding in some of them with EH results [24]. Karplus et al. used EH and density functional self-consistent field-multiple scattering-Xα (Xα) calculations to examine the electronic structure of molecular cubic clusters containing a $Co_8(\mu_4\text{-}S)_6$ core [25]. More recently, Rösch et al. performed DFT calculations in order to analyze the metal-metal bonding in $Ni_8(\mu_4\text{-}PPh)_6(CO)_8$ [26]. We ourselves recently rationalized the structural arrangement of the whole set of compounds listed in Table 1 with the aid of EH and Xα calculations [27].

2.1 The 120-Electron Cubic Species $M_8(\mu_4\text{-}E)_6L_8$

The electronic structure of the 120-electron cubic system $M_8(\mu_4\text{-}E)_6L_8$ has often been described in terms of a localized bonding scheme, i.e. represented by a Lewis structure with 2-center/2-electron bonds, in which the M-M bonding is considered as "cubane-like". Within such a bonding model, the transition metal atoms obey the 18-electron rule, but the main-group E atoms, being pentavalent, do not follow the octet rule. Localized Lewis formulae are inadequate in describing properly the bonding in hypervalent molecules [28]. Indeed, each capping main-group atom or group of atoms, e.g., PR in **1**, possesses only three frontier molecular orbitals (FMO) to ensure the bonding with the metallic square to which it is tethered. It follows that only a delocalized MO description of the 120-electron cubic species can account for the nature ("cubane-like" or not) of the M-M bonding. We have shown that such a delocalized MO picture could easily be obtained by deriving the MO diagram of the 120-electron species $Ni_8(\mu_4\text{-}E)_6L_8$ with O_h symmetry from the interaction of the frontier molecular orbitals (FMO) of a $[Ni_8L_8]^{12+}$ cube with the FMOs of a capping $[E_6]^{12-}$ octahedron (see Fig. 1) [27].

The FMOs of the $[Ni_8L_8]^{12+}$ cubic unit can be obtained qualitatively by combining the FMOs of the eight Ni-L fragments. Such a fragment, as shown on the left-hand side of Fig. 1, presents a block of five d-type orbitals (one, two, and two of σ, π and δ symmetry, respectively) below a radial σ-type sp hybrid and a set of two tangential π-type orbitals of Ni 4p parentage [29]. One can deduce from group theory that the three upper diffuse FMOs (sp_σ and p_π) of the eight Ni-L units will generate a set of 12 bonding MOs, largely separated from a set of 12 antibonding MOs. Similarly, the combination of the three more contracted lower d_σ and d_π orbitals will lead to the formation of a set of 12 bonding and a set of 12 antibonding levels. The splitting between these two sets of MOs is expected to be rather small because of the poor d-overlap. The two δ-type FMOs are expected to interact even more weakly, giving rise to a block of 16 nearly nonbonding MOs. The resulting level ordering for the $[Ni_8L_8]^{12+}$ moiety is given on the left hand-side of Fig. 1.

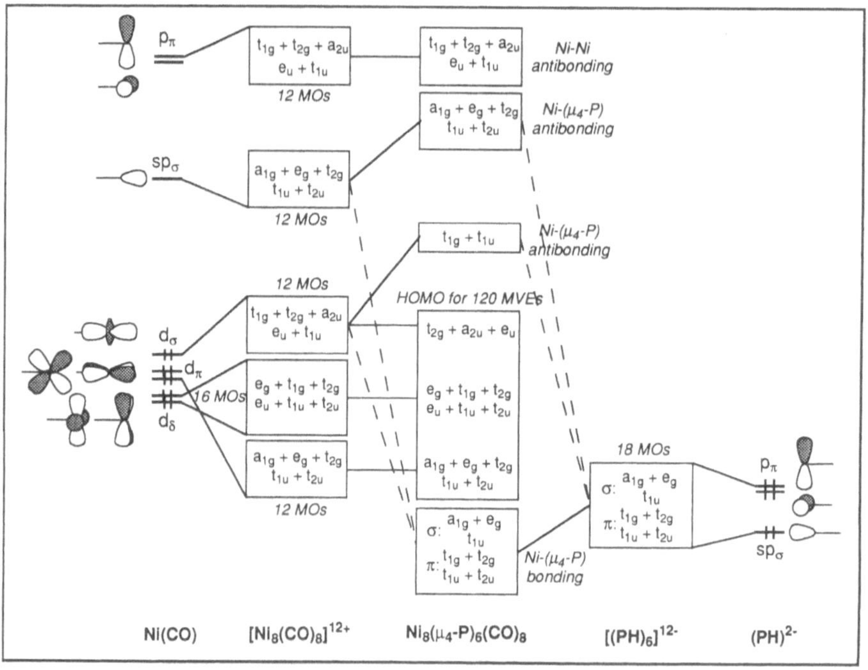

Fig. 1. Qualitative MO diagram for the model $Ni_8(\mu_4\text{-}PH)_6(CO)_8$ obtained from the interaction of the cubic $[Ni_8(CO)_8]^{12+}$ and octahedral $[(PH)_6]^{12-}$ fragments. The FMOs of the Ni(CO) and PH moieties are shown on the *far left-* and *far right-hand side*, respectively

The three FMOs of the six noninteracting E groups constituting the octahedral $[E_6]^{12-}$ moiety combine with each other to give a set of 18 nonbonding MOs (6 of σ-type and 12 of π-type) represented on the right-hand side of Fig. 1.

The major bonding interactions (i.e., 2-electron/2-orbital) between the two fragments $[Ni_8L_8]^{12+}$ and $[E_6]^{12-}$ are expected to occur between the 18 formally occupied FMOs associated with $[E_6]^{12-}$ and 18 lowest unoccupied FMOs of $[Ni_8L_8]^{12+}$ of the same symmetry. Among these vacant metallic FMOs, only 6 can be picked up in the upper part of the d block (the 34 remaining ones are occupied for the formal charge of 12+). The 12 other FMOs come from the Ni-Ni (s+p) bonding block, which constitutes a set of low-lying and diffuse acceptor orbitals. For symmetry reasons, the six d-type metallic orbitals involved in the interaction are necessarily of t_{1g} and t_{1u} symmetry [27]. The resulting qualitative diagram of the $[Ni_8(\mu_4\text{-}E)_6L_8]$ species is illustrated in the middle of Fig. 1.

Although the destabilizing 4-electron-2-orbital interactions and second-order mixings are not taken into account, we have shown that such a simplified diagram provides a general understanding of the chemical bonding in the cubic clusters considered [27]. The level ordering of the MO diagram of the $M_8(\mu_4\text{-}E)_6L_8$ species is expected to be the following: in addition to a set of 8 low-lying M-L bonding MOs (not indicated in Fig. 1), a set of 18 M-E bonding MOs will be found at lower energy, corresponding to the 24 M-(μ_4-E) bonding contacts (note the delocalized bonding picture here). Higher in energy, a block of 34 MOs mainly metallic in character is found, separated from M-E antibonding orbitals by a rather large energy gap. A closed-shell count of 120 MVEs is obtained when this whole d-block is filled up.

The calculations of Burdett and Miller, and our own are in agreement with these qualitative considerations, showing a HOMO-LUMO gap larger than 1 eV for the count of 120 MVEs [24, 27]. The M-(μ_4-E) bonding is maximized for the 120-MVE count [27]. On the other hand, although a rather strong M-M bonding is present in these species, mainly due to through-space M-M interactions but also via through-bond M-E interactions, the M-M bonding is not at its maximal value. Indeed, the top of the d band is weakly antibonding, suggesting that the electron count of these cubic M_8E_6 compounds is primarily governed by the M-E rather than by the M-M interactions [24, 27].

2.2
Fewer Electrons

Among the $M_8(\mu_4\text{-}E)_6L_8$ species listed in Table 1, none has more than 120 MVEs. Electron counts larger than 120 MVEs would result in the population of MOs having a strongly antibonding character, which would induce some instability. On the other hand, a certain number of them have less than 120 MVEs. Clearly, the number of valence electrons that maximizes the cluster stability of these species depends upon the nature of the metal atoms and the terminal ligand system. The optimal 120-MVE count is favored with electronegative metals and/or π-acceptor terminal ligands. Lower electron counts are observed if Fe or Co is substituted for Ni, or if π-donor terminal ligands are present instead of π-acceptor ligands.

Calculations that we [27] and Burdett and Miller [24] carried out on different models have indicated that, since the top of the d band is weakly antibonding, the M-M bonding is strengthened upon its depopulation. Electron counts lower than 120 MVEs and open-shell configurations are then possible for clusters bearing terminal π-donor ligands.

The possible depopulation of the top of the metallic d band, which is entirely delocalized over the cube, allows a large range of electron counts (from 120 to 99 so far) without altering the cubic metallic architecture. In agreement with this statement, the M_8E_6 core of the electron-deficient clusters given in Table 1 is never far from O_h symmetry regardless of their electron counts, and some of them are paramagnetic [8]. The moderate M-M antibonding character noted for some MOs which might be depopulated is confirmed for instance by the weak diminution of the Ni-Ni separations in the electron-deficient nickel species compared to that of the 120-electron compounds (see Table 1). Xα calculations performed by Karplus and coworkers have shown that singlet-state and triplet-state electron configurations in the electron-deficient cubic $Co_8(\mu_4\text{-}S)_6$ species with an even number of MVEs are nearly isoenergetic and that the distortion away from the cubic O_h symmetry was very weak [25]. The same conclusions were reached for the $Fe_8(\mu_4\text{-}S)_6$ clusters [27]. It appears that the important connectivity between the atoms of the cluster core does not allow them to distort significantly away from the hexacapped cubic architecture. This opposition to a Jahn-Teller distortion is reminiscent of solid state chemistry.

The count of 76 MVEs is likely to constitute the lowest hypothetical limit for these cubic species, which preserves the M-M cubane-type bonding mode. Cubic architectures such as $Nb_8(\mu_4\text{-}PR)_6(PR'_3)_8$ or $[Mo_8(\mu_4\text{-}GeR)_6Cl_8]^{2-}$ containing electron-poor metal atoms and possessing diffuse atomic orbitals, have been predicted [27].

2.3 Fewer Terminal Ligands

All the compounds such as **2**, having an incomplete terminal ligand shell, possess less than 120 MVEs (see Table 1). It appears that when terminal ligands borne by the metal atoms are missing, the expected close-shell electron count for compounds of formula $M_8(\mu_4\text{–}E)_6L_n$ is 120–2 × (8 – n) MVEs [27]. Indeed, in cubic $M_8(\mu_4\text{–}E)_6L_n$ species, M-L bonds result from the combination between a metallic sp hybrid and the σ lone pair orbital of L. When n terminal ligands are lost, the corresponding metallic sp hybrids become nonbonding but remain high in energy and cannot be occupied, leading to an electron count diminished by 2n units compared to that of the $M_8(\mu_4\text{–}E)_6L_8$ parent species (see Fig. 2). This situation is reminiscent of the stability of 16-electron square-planar ML_4 complexes. This electron count occurs for electronegative metal atoms and terminal π-acceptor ligands and might be less apparent with less electronegative metal atoms and/or terminal π-donor ligands.

The coordination deficiency of some Ni centers in the $M_8(\mu_4\text{–}E)_6L_n$ species leads to a weak distortion of the metallic cube. Additional Ni-Ni bonding occurs

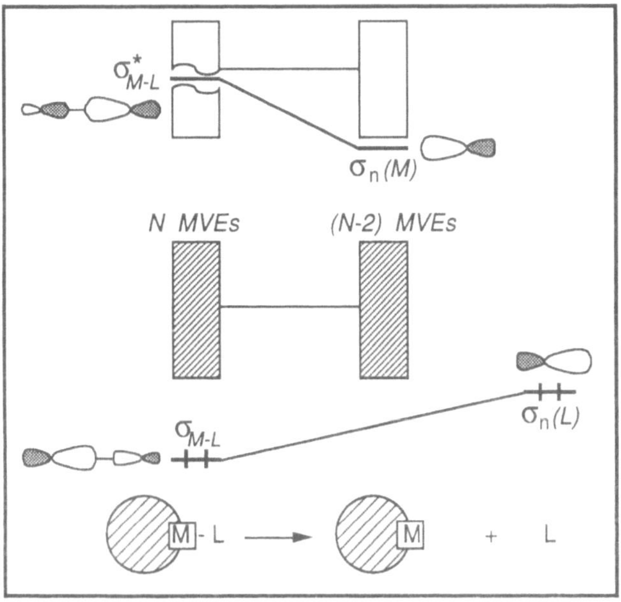

Fig. 2. Electronic effect of the loss of a terminal ligand in an $M_8(\mu_4\text{-}E)_6L_n$ cluster species

between nonadjacent Ni atoms to which no terminal ligand is attached. The origin of this attractive interaction between nonbonded metals is due to second order mixing of in-phase combination of the free sp Ni hybrids into occupied levels of proper symmetry. DFT results obtained by Rösch et al. on different ligated $M_8(\mu_4\text{-}E)_6L_n$ have indicated a direct relationship between the number of terminal ligands and the M-M bonds [26]. The cluster loses its metallic character as the number of terminal ligands increases, leading to a substantial elongation of the M-M separations.

2.4
More Terminal Ligands

The black octanuclear μ_4-phosphinidene cluster of rhodium $Rh_8(\mu_4\text{-}P(C_5H_9))_6(CO)_9$ (**3**) [23] is strongly related to the cubic $M_8(\mu_4\text{-}E)_6L_8$ compounds. As in $Ni_8(\mu_4\text{-}PPh)_6(CO)_8$, the cluster core consists of a central M_8 cube capped on each face by a phosphinidene ligand. However, the presence of an extra carbonyl ligand tethered to one of the Rh atoms in 3 induces some distortion of the metal cage and prevents it from being fully isostructural to compound 1. With 114 MVEs, cluster 3 is related to the electron-deficient complexes listed in Table 1.

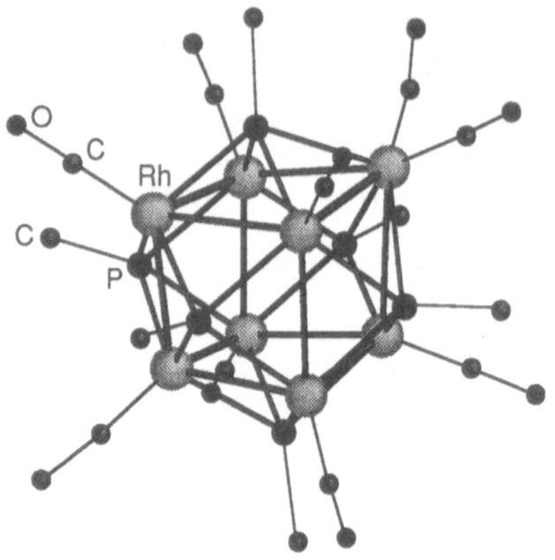

$Rh_8(\mu_4\text{-}P(C_5H_9))_6(CO)_9$

3

3 Other Cubic M_8 Clusters

3.1 Large Clusters Containing Cubic M_8 Units

The crystal structure of $Pd_{20}(\mu_5\text{-}As)_{12}(PPh_3)_{12}$ (**4**) is formed from a regular Pd_8 cube, the edges of which are bridged by a μ-$Pd(PPh_3)$ unit [30]. In addition, μ_5-As bridging ligands come to complete the sheath of the central metal cube. This compound possesses 260 MVEs, that is 28 electrons less than predicted by the

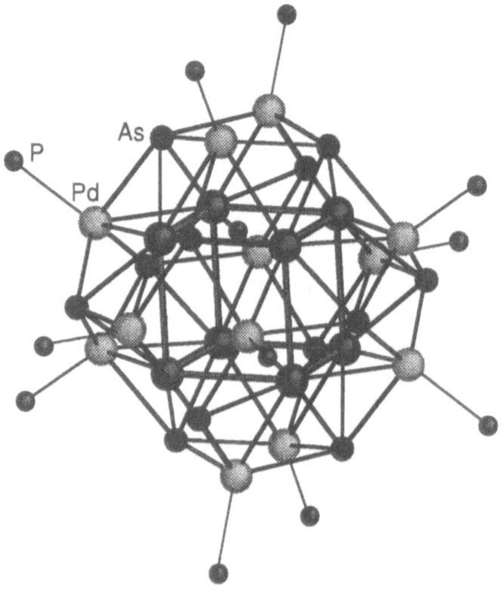

$Pd_{20}(\mu_5\text{-}As)_{12}(PPh_3)_{12}$

4

18-electron rule, but 12 electrons more than expected from the 16-electron rule. Assuming a formal division of **4** into an internal M_8 cube and an external 12-center polyhedron, application of the Mingos' principle of *inclusion* [31] leads to 264 ($12n_s + \Delta_i = 12 \times 12 + 120$), which is only 4 electrons more than that counted in **4**.

3.2 Copper Species

Fackler and coworkers synthesized copper cubic cluster salts such as $[(NPhMe_3)_4Cu_8(i\text{-MNT})_6]$ (i-MNT$=S_2CC(CN)_2$) (**5**) [32], $[(PPh_4)_4Cu_8(DTS)_6]$ ($DTS=S_2C_4O_2$) [33], and $[(PPh_4)_4Cu_8(DED)_6]$ ($DED=S_2CC(COEt)_2$) over 25 years ago [34]. These compounds exhibit a cubic $[Cu_8(\mu\text{-}S)_{12}]^{4-}$ core of T_h symmetry with the S atoms of the bidendate ligands bridging the edges of the cube. They are characterized by a count of 128 MVEs. Assuming a formal charge of 2- per chelate ligand, each Cu atom has a formal oxidation state of 1+, corresponding to 16-electron Cu centers. Disregarding the rather short Cu-Cu separations measured experimentally (ca. 2.85 Å), this is a stable geometry for triangular-planar CuL_3 fragments. On the other hand, 18-electron Cu centers would require a Cu-Cu bond order of 2/3. A possible explanation for the short Cu-Cu contacts would be d^{10}-d^{10} bonding through second order mixing of vacant 4s and 4p Cu orbitals in the occupied 3d block. This is not in agreement with theoretical calculations which indicate no net bonding between the Cu atoms, but strong Cu-S bonding [35].

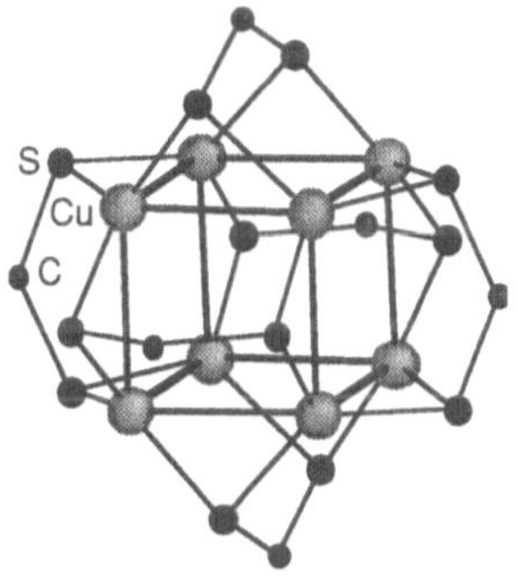

$[Cu_8(i\text{-MNT})_6]^{4-}$

5

3.3 Metallo-Carbohedrene Clusters

Based on an omnicapped distorted metallic cube, molecular metallocarbohedrene $M_8(C_2)_6$ clusters need to be mentioned here. Initially reported by Castelman several years ago for M=Ti, they are now observed for different early transition metals [36]. Numerous theoretical investigations, sometimes controversial, have been devoted to this kind of electron-deficient compound [37]. Recent ab initio post-HF calculations carried out by Bénard et al. suggest an $M_8(C_2)_6$ cage of T_d symmetry made of a "small" metallic tetrahedron surrounded by a "large" metallic tetrahedron rather than a regular dodecahedral arrangement of T_h symmetry, as initially thought [37]. The probable ground state of the 44-MVE $Ti_8(C_2)_6$ species is a localized state with four antiferromagnetically coupled d electrons accommodated in the small metallic tetrahedron. The ground state of the more electron-rich 52-MVE $V_8(C_2)_6$ and $Nb_8(C_2)_6$ compounds is also characterized by four unpaired electrons either localized on the small metallic tetrahedron and antiferromagnetically coupled, or localized on the large metallic tetrahedron and displaying ferromagnetic coupling [37].

4
Metal-Centered Hexacapped $M_9(\mu_4\text{-E})_6L_8$ Species

Recent efforts to incorporate elements into the metallic cube of these clusters have given rise to a new class of electron-rich compounds of formula $M_9(\mu_4\text{-E})_6L_8$, where a metal atom is encapsulated at the center of the metallic cubic core. The 124-MVE compound $Ni_9(\mu_4\text{-GeEt})_6(CO)_8$ (**6**) characterized by Dahl and coworkers provides a specific example [38]. Several metal-centered hexacapped cubic $M_9(\mu_4\text{-E})_6L_8$ clusters analogous to 6 have been structurally characterized. These compounds, summarized in Table 2, are known only for M=Ni or Pd so far. Exhibiting MVE counts varying from 121 to 130, they are more electron-rich that their noncentered $M_8(\mu_4\text{-E})_6L_8$ parents, which possess a maximum of 120 MVEs (see above).

Few theoretical studies have been made on this family of compounds. Wheeler has analyzed the electronic structure of $Ni_9(\mu_4\text{-Te})_6(PEt_3)_8$ with EH cal-

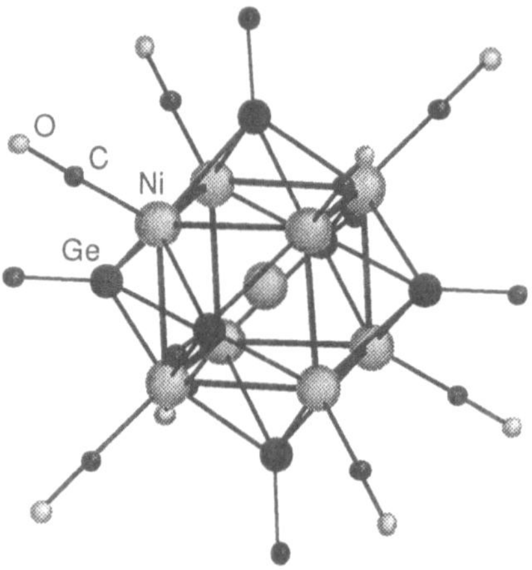

$Ni_9(\mu_4\text{-GeEt})_6(CO)_8$

6

Table 2. Metal-centered molecular cubic $M_9(\mu_4\text{-}E)_6L_8$ clusters characterized by X-ray diffraction

Compound	$d(M_s\text{-}M_s)^a$	$d(M_i\text{-}E)/d(M_s\text{-}E)^b$	MVE^c	Ref.
$Ni_9(\mu_4\text{-}Te)_6(PEt_3)_8$	2.86	1.17	130	39
$Pd_9(\mu_4\text{-}As)_6(PPh_3)_8$	3.11	1.09	124	30
$Pd_9(\mu_4\text{-}Sb)_6(PPh_3)_8$	3.26	1.10	124	40
$Ni_9(\mu_4\text{-}GeEt)_6(CO)_8$	2.67	1.17	124	38
$Ni_9(\mu_4\text{-}As)_6(Pn\text{-}Bu_3)_8$	2.85	1.10	124	41
$Ni_9(\mu_4\text{-}Sb)_6(PPh_3)_8$	2.99	1.11	124	41
$Ni_9(\mu_4\text{-}Bi)_6(PPh_3)_8$	3.00	1.15	124	41
$Ni_9(\mu_4\text{-}P)_6(PCy_3)_6Cl_2$	2.80	1.10	122	40
$Ni_9(\mu_4\text{-}As)_6(PPh_3)_6Cl_2$	2.81	1.13	122	42
$Ni_9(\mu_4\text{-}As)_6(PPh_3)_5Cl_3$	2.81	1.13	121	41

[a] Averaged surface metal-surface metal distance (Å).
[b] Ratio between interstitial metal-capping atom and surface metal-capping atom distances.
[c] Total valence metallic electron count.

culations [43]. Hoffmann and coworkers have also used EH calculations to examine the electronic structure of this molecular cubic cluster in order to draw relationships with nickel-tellurium extended structures [44]. More recently, we have carried out EH and Xα calculations to investigate the bonding in these clusters as a function of different parameters, such as the electron count or the nature and the size of the different elements constituting the cluster cage [45].

4.1 Closed-Shell Electronic Configurations

We have shown that a localized two-center-two-electron bonding scheme on which the 18-electron rule is based cannot be applied to the cubic $M_9(\mu_4\text{-}E)_6L_8$ species [45]. A delocalized approach is then necessary to describe the metallic bonding mode in these compounds. According to the inclusion principle developed by Mingos and coworkers [31], the M_9 compounds given in Table 2 should possess either 114 or 120 MVEs. The actual electron counts observed for the cubic clusters containing an encapsulated transition-metal atom show that these electron-counting rules may not be reliably applied to this class of compound.

The qualitative electronic structure of the $M_9(\mu_4\text{-}E)_6L_8$ compounds can be described as resulting from the interaction between the interstitial metal atom M_i and its $M_8(\mu_4\text{-}E)_6L_8$ host [45]. The nine AOs of the interstitial metal atom span a_{1g} (s) + t_{1u} (x, y, z) + e_g ($x^2\text{-}y^2$, z^2) + t_{2g} (xy, xz, yz) under O_h symmetry. Strong bonding interactions are expected between its high-lying diffuse s (a_{1g}) and p (t_{1u}) AOs and some corresponding levels of the d-block of the metallic M_8 cage. As a consequence, the high-lying s and p AOs of M_i are strongly destabilized and cannot be populated. Therefore, supposing a 120-MVE $M_8(\mu_4\text{-}E)_6L_8$ cage having a significant HOMO-LUMO gap (see above), these particular a_{1g} and t_{1u} interactions which occur when M_i is incorporated at its center are not expected to change this favored electron count.

Such a result is not so straightforward for the interactions involving the low-lying and more contracted d AOs of M_i. We have shown that four different cases leading to four different closed-shell MVE counts are possible a priori, depending upon the strength of the interaction of the M_i d orbitals with corresponding FMOs of the cubic metallic host [45]. They are shown in Fig. 3. If both e_g and t_{2g} AOs of M_i interact strongly with some corresponding metallic levels of the cage, the resulting out-of-phase combinations will be sufficiently antibonding to lie at high energy and will not be occupied. The closed-shell electron count of 120 then remains unchanged (electron configuration noted {120}, see Fig. 3a) in agreement with Mingos' rules. If both the e_g and t_{2g} orbitals interact weakly with the metallic cage, the out-of-phase combinations will remain at a relatively low energy and therefore will be occupied (see Fig. 3b). This increases by five the number of levels in the cluster d-block, leading to a favored closed-shell MVE count of 120 + 10 = 130, corresponding to the $\{120\}(e_g)^4(t_{2g})^6$ configuration. With a strong t_{2g} interaction and a weak e_g interaction, the favored closed-shell electron count is 120 + 4 = 124 (configuration $\{120\}(e_g)^4$, see Fig. 3c). Finally, if the e_g interaction is strong and the t_{2g} interaction is weak, the favored closed-shell electron count is 120 + 6 = 126 (configuration $\{120\}(t_{2g})^6$, see Fig. 3d). The strength of the M_s-M_i and M_s-M_s bonds will depend on the electron count.

This qualitative approach should result in a simple classification scheme. This is not actually the case [45]. There is only one example among the characterized M_9 compounds, namely $Ni_9(\mu_4\text{-}GeEt)_6(CO)_8$ [38], the electronic structure of which fits with one of the electronic structures shown in Fig. 3. Indeed the EH and Xα calculations on the 124-electron model $Ni_9(\mu_4\text{-}GeH)_6(CO)_8$ show that it corresponds to the situation depicted in Fig. 3c. Examination of the EH bond overlap populations indicate that neither the M-M nor the M-E bonding contacts are maximized for this count of 124 MVEs [45]. The computed EH and Xα electronic populations on the central nickel atom reflect the rather strong t_{2g} interaction. Conversely, the e_g population is close to 4, confirming that these orbitals do not play any significant role in the Ni_s-Ni_i bonding.

We concluded from our analysis that, amid the four closed-shell configurations proposed in Fig. 3, the two involving a weak t_{2g} interaction between the encapsulated metal atom and the cubic unit, which correspond to the 126 and 130-MVE counts, are unlikely to exist [45]. Indeed, the corresponding overlap is always significant, the t_{2g} AOs of M_i pointing towards the corners of the metallic cube. Consequently, it appears impossible to cancel fully the antibonding character of the t_{2g} out-of-phase combination. Even when the M-M distances are rather long, like in $Pd_9(\mu_4\text{-}Sb)_6(PPh_3)_8$, this level is still somewhat antibonding and lies in the middle of an energy gap. This situation favors its vacancy or it partial occupation (see below). On the other hand, 120-MVE M_9 compounds for which both e_g and t_{2g} interactions are strong should be attainable with transition metals less electron-rich than Ni and small-size E capping ligands [45].

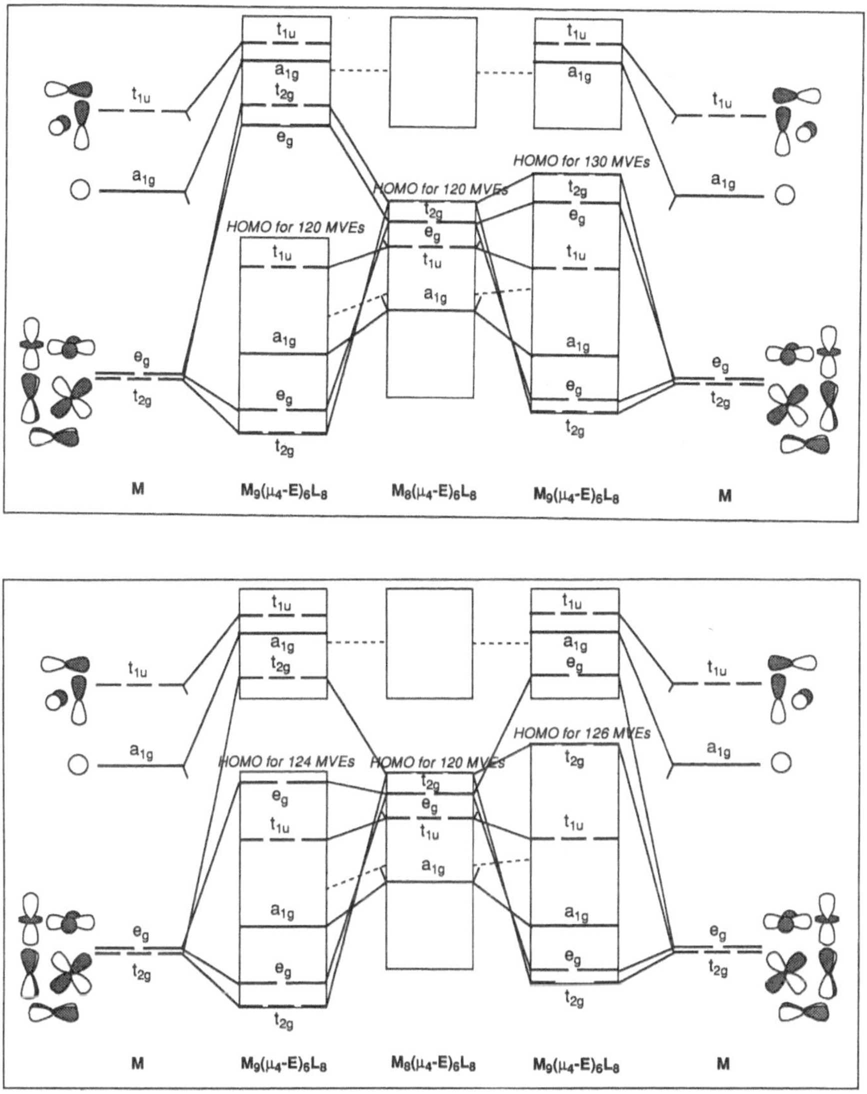

c d

Fig. 3a–d. Qualitative MO diagrams predicted for a cluster $M_9(\mu_4\text{-}E)_6L_8$ depending on the magnitude of the interaction of the central metal atom with its metallic host

4.2
Alternative Open-Shell Electronic Configurations

Exploration of the electronic structure of the other compounds listed in Table 2 has indicated the preference for open-shell ground-state configurations [45]. For instance, the ground-state electron distribution of the 124-MVE clusters

$Pd_9(\mu_4\text{-}E)_6(PPh_3)_8$ (E=As, Sb) [30, 40] and $Ni_9(\mu_4\text{-}E)_6(PR_3)_8$ (E=As, Sb, Bi, R=P*n*-Bu_3, PPh_3) [41] corresponds to $\{120\}(t_{2g})^4(e_g)^0$. It is different from the closed-shell configuration $\{120\}(e_g)^4(t_{2g})^0$ computed for the isoelectronic 124-MVE $Ni_9(\mu_4\text{-}GeEt)_6(CO)_8$ species. We have analyzed the parameters which may be responsible for this difference of electronic configurations [45]. Contrary to a superficial analysis, the difference is not due to the magnitude of the e_g interaction between the interstitial metal atom and the cubic cage. We have found that this interaction is weak in all cases. It turns out that in this case the crucial parameter is the particular nature of the capping ligands. Electron counts of 124 MVEs corresponding to open-shell configurations are favored when the capping E ligands are strong donors and/or when they are bare. For instance, calculations performed on the model $Ni_9(\mu_4\text{-}GeH)_6(CO)_8$ indicate a closed-shell configuration $\{120\}(e_g)^4(t_{2g})^0$, whereas calculations carried out on the isoelectronic model $[Ni_9(\mu_4\text{-}Ge)_6(CO)_8]^{6-}$ show an open-shell configuration $\{120\}(t_{2g})^4(e_g)^0$, similar to that of $Pd_9(\mu_4\text{-}E)_6(PPh_3)_8$. This is due to the energy difference of the σ-type FOs of E vs those of E-R [45]. With bare capping ligands, the levels which may be partly populated are of t_{2g} and (for large electron counts) e_g and t_{1g} symmetry. This is the case of the 130-MVE cluster $Ni_9(\mu_4\text{-}Te)_6(PEt_3)_8$ [39]. An open-shell configuration $\{120\}(e_g)^4(t_{1g})^4(t_{2g})^2$ is computed for this compound [45]. The configuration $\{120\}(e_g)^0(t_{2g})^6$ is mostly less stable suggesting that, as mentioned above, too many electrons in the antibonding t_{2g} MO induce an important loss of bonding between the central and surface metal atoms, rendering the cluster unstable with respect to dissociation.

It appears that metal-centered $M_9(\mu_4\text{-}E)_6$ compounds with open-shell electronic configurations are most commonly observed. Nevertheless, these open-shell configurations never lead to strong Jahn-Teller distortion [30, 39–42]. A nearly regular cubic architecture is observed for these particular clusters. This is mainly due to the high connectivity of the different atoms constituting the cluster. Making new bonds induces a lengthening or a breaking of other bonds. As stated above, this situation is reminiscent of that often encountered in solid state chemistry, for body-centered-cubic metals for instance.

5
Distorted Cubic M_9 Compounds

Highly distorted metal-centered M_9 cubic architectures may be encountered, however, when the ligand environment of the metallic cage is nonsymmetrical. This is the case for instance of the related species $[Co_9Bi_4(CO)_{16}]^{2-}$ (**7**), recently characterized by Eveland and Whitmire [46]. Being viewed as a metal-centered elongated Co_8 cube with four Bi atoms capping the four rectangular faces, cluster **7** bears some structural resemblance to the hexacapped cubic $M_9(\mu_4\text{-}E)_6$ species such as $Ni_9(\mu_4\text{-}Bi)_6(PPh_3)_8$ mentioned above [30, 38–42]. In both cases, the central metal atom is twelve-coordinated. However, there are twelve Ni_s-Ni_s bonds and no direct Bi-Ni_i interactions in $Ni_9(\mu_4\text{-}Bi)_6(PPh_3)_8$ whereas there are only eight Co_s-Co_s bonds and four additional Bi-Co_i bonds in **7**. We have seen that the cubic compounds $M_9(\mu_4\text{-}E)_6L_8$, known only for M=Ni or Pd so far, have electron counts varying from 121 to 130 [45]. With 127 MVEs, cluster **7** is a new

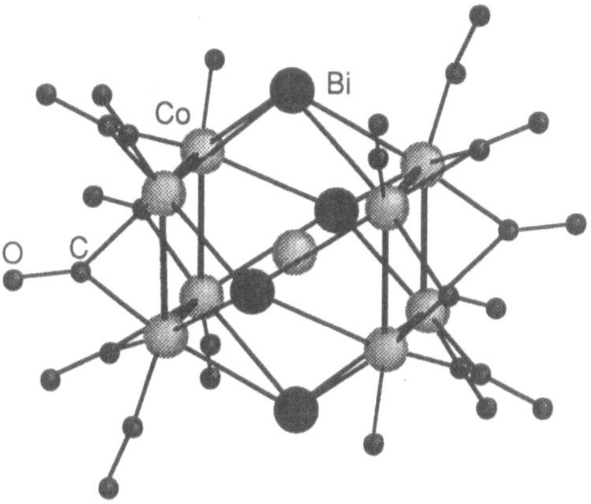

$[Co_9(\mu_5\text{-}Bi)_4(\mu\text{-}CO)_8(CO)_8]^{2-}$

7

example extending this class of compounds. We have recently carried out EH and DFT calculations in order to provide a comparison with the $M_9(\mu_4\text{-}E)_6L_8$ cubic clusters [47].

The electronic structure of cluster 7 can be described as resulting from the interaction between the interstitial Co atom and its elongated cubic host $Co_8Bi_4(CO)_{16}$. The MO diagram of the $Co_8Bi_4(CO)_{16}$ fragment itself can be derived from that of a regular tetracapped cube (TCC) [48]. Although the symmetry of the TCC $Co_8(\mu_4\text{-}Bi)_4(CO)_8(\mu\text{-}CO)_8$ fragment is D_{4h}, EH calculations indicate an MO diagram strongly similar to that of the O_h cubic clusters $M_8(\mu_4\text{-}E)_6L_8$ [27]. Particularly, a significant HOMO-LUMO gap is computed for the count of 120 MVEs. Upon stretching along the four-fold axis, only three MOs are significantly stabilized, leading to a closed-shell MVE count of 126, rather than 128 as expected upon breaking four M-M localized bonds [47]. Once the interstitial Co_i atom sits at the center of the the elongated TCC fragment $Co_8(\mu_4\text{-}Bi)_4(CO)_8(\mu\text{-}CO)_8$, one is left with a situation similar to that depicted in Fig. 3b, where only the 4s and 4p AOs of Co_i interact significantly with the elongated TCC host. This leads to the hypothetical closed-shell MVE count of 136. Such clusters should exist with the appropriate transition metal.

Electron counts lower than 136 would correspond to the depopulation of the upper d-block MOs, which are slightly Co_s-Co_i and Bi-Co_i antibonding, strengthening the bonding of Co_i with the Co_8Bi_4 fragment. This is what happens for compound 7. With a count of 127 MVEs, DFT calculations indicate an open-shell electron configuration [122] $(b_{2g})^2(e_g)^1(a_{1g})^2$ for the paramagnetic species 7 [47].

An alternative distortion of an body-centered M_9 cube, of "D_{2d}" type is observed in the cluster $Pd_9(\mu{:}\eta^5,\eta^2\text{-}As_2)_4(PPh_3)_8$ (**8**) [40]. Counting the $(\mu{:}\eta^5, \eta^2\text{-}As_2)$ units which asymmetrically bridge four faces of the metallic cube as 6-electron donor ligands, compound 8 contains 130 MVEs.

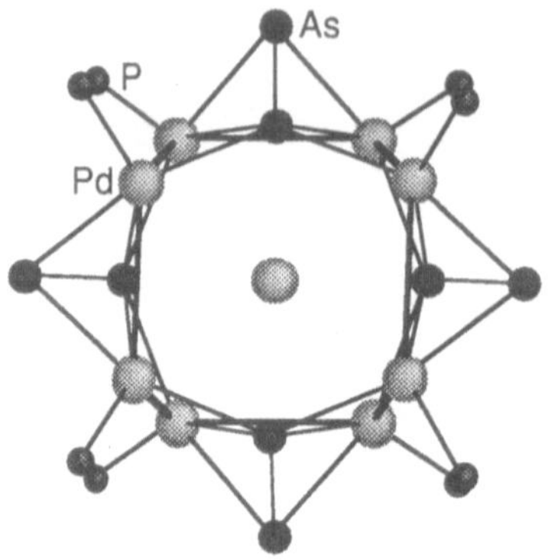

$Pd_9(\mu{:}\eta^5,\eta^2\text{-}As_2)_4(PPh_3)_8$

8

6
Incorporation of Main Group Atoms in Cubic $M_8(\mu_4\text{-}E)_6L_8$ Cavities

A few years ago, Wheeler examined theoretically the possibility of encapsulating main-group atoms at the center of the metallic $M_8(\mu_4\text{-}E)_6L_8$ cube [43]. Using EH calculations carried out on the model $Ni_8(\mu_8\text{-}Te)(\mu_4\text{-}Te)_6(H)_8$, he predicted two favorable MVE closed-shell electron counts, 110 and 126, for these hypothetical cubic species. Though these electron counts are in disagreement with the inclusion principle [31] which predicts 114 or 120 MVEs, the strong bonding of the interstitial Te atom with both Ni and Te atoms of the framework computed for the 126-MVE count led him to propose that such $Ni_8(\mu_8\text{-}Te)$ species should be stable. He pointed out however, that the close grouping of MOs in the HOMO region, as well as the variety of electron counts observed for the empty cubic $M_8(\mu_4\text{-}E)_6L_8$ clusters (see above), may imply the possibility of other electron counts depending on details of the ligands [43].

Some years later, the synthesis of the 119-MVE compound $Ni_8(\mu_8\text{-}As)(\mu_4\text{-}As)_6(PEt_3)_8$ (**9**) was reported by Fenske and Vogt [41]. This is the only example of cubic $M_8(\mu_4\text{-}E)_6L_8$ species known so far with a main group atom at the center of a cubic $M_8(\mu_4\text{-}E)_6L_8$ cage.

With this compound in mind, we have recently analyzed the bonding in the cubic clusters $M_8(\mu_8\text{-}E')(\mu_4\text{-}E)_6L_8$ with the help of EH and DFT calculations [49]. The conclusions are slightly different from those drawn by Wheeler since two possible closed-shell electron counts are found, namely 120 and 122, depending on the nature of the incorporated main group element.

As for the metal-centered $M_9(\mu_4\text{-}E)_6L_8$ species, the bonding in the cluster $M_8(\mu_8\text{-}E')(\mu_4\text{-}E)_6L_8$ can be envisioned from the interaction of the central E′ atom with the metallic $M_8(\mu_4\text{-}E)_6L_8$ cubic unit. Two favorable closed-shell electron configurations are expected. In simple terms, one results when both the s and p AOs of E′ (which span a_{1g} and t_{1u} in O_h symmetry, respectively) interact

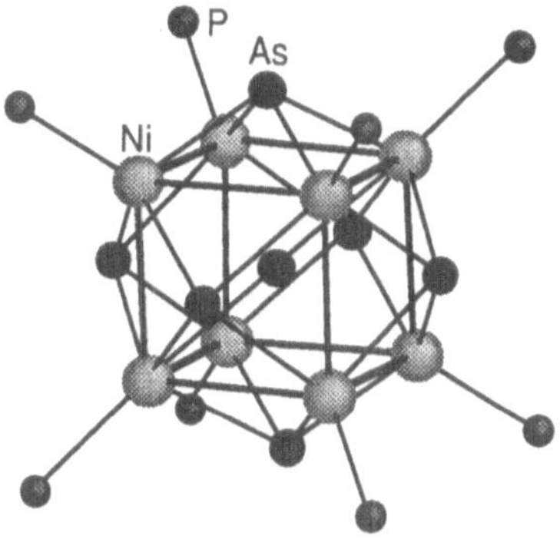

$Ni_8(\mu_8\text{-}As)(\mu_4\text{-}As)_6(PPh_3)_8$

9

strongly with metallic MO counterparts of the cubic host. This situation, which leads to the expected closed-shell count of 120 MVEs as the empty parent molecule, is illustrated in Fig. 4a. With heavy main group elements, however, the s AO of E′ is rather contracted and low in energy. Consequently, its interaction with a_{1g} MOs of the cubic unit is weaker, leading to an out-of-phase combination which remains sufficiently low in energy to be occupied. This favors a closed-shell count of 122 MVEs (see Fig. 4b). DFT calculations carried out on different

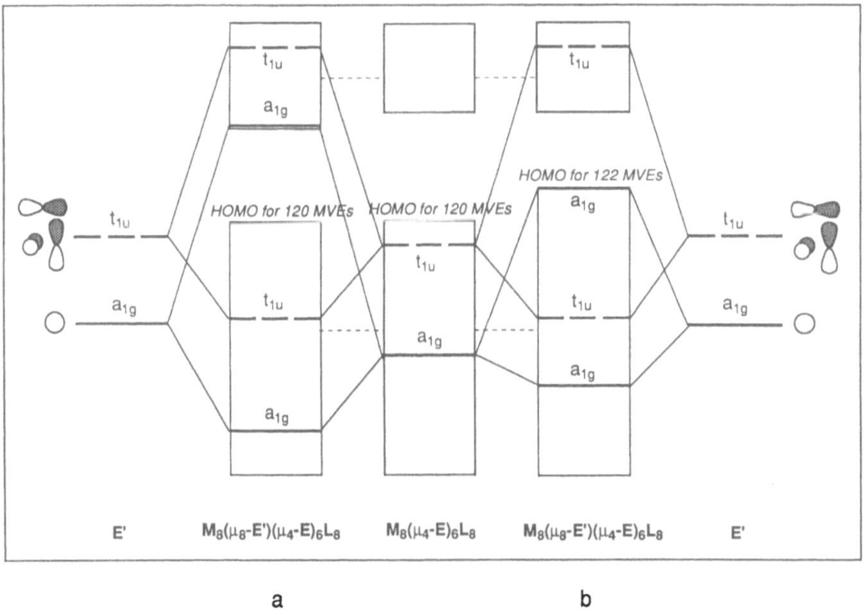

Fig. 4a, b. Qualitative MO diagrams predicted for a cluster $M_8(\mu_8\text{-}E')(\mu_4\text{-}E)_6L_8$ depending on the magnitude of the interaction of the central main group atom E′ with its metallic host

$M_8(\mu_8\text{-}E')(\mu_4\text{-}E)_6L_8$ indicate that the 120-MVE count is favored with small-size E′ atoms such as phosphorus or sulfur, whereas the 122-MVE count is preferred with large-size E′ atoms such as As or Te [49].

In view of the large variety of electron counts encountered in empty and metal-centered cubic compounds, Wheeler [43] and we [45] pointed out that main group atom-centered clusters based on the same architectural unit should be observed with open-shell electron configurations. This is demonstrated with Fenske's cluster **9**. DFT calculations give the open-shell electronic configuration $\{116\}(a_{1g})^2(e_g)^1$ as the ground-state configuration for the 119-MVE As-containing species **9** [49]. The occupation of the slightly Ni_s-As_i antibonding a_{1g} cluster MO is in agreement with the rather long Ni_s-Ni_s bonds compared to the Ni-Ni bonds measured in the 124-MVE metal-centered cluster $Ni_9(\mu_4\text{-GeEt})_6(CO)_8$ for instance (2.89 Å vs. 2.67 Å) [38].

7
Other Main Group Atom-Centered Cubic Species

$Cu_{20}(\mu_8\text{-Se})Se_{12}(PEt_3)_{12}$ (**10**) and $Cu_8(\mu_8\text{-S})(S_2P[O\textit{i}\text{-}C_3H_7]_2)_6$ (**11**) provide examples of copper clusters containing chalcogen atoms encapsulated in a metallic cubic array. The former, characterized by Fenske and Krautscheid [50], adopts a spherical arrangement with a rather large Cu_8 (μ_8-Se) unit at the center (Cu-Cu distances of ca. 3 Å). With 298 MVEs, the 18-electron rule is exceeded by only two electrons, assuming that all Cu-Cu separations under 2.90 Å are bonding [50]. The bonding situation is then different from that in the structur-

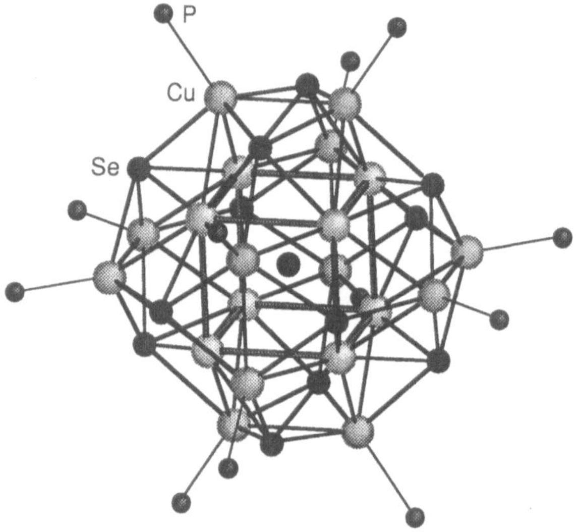

$Cu_{20}(\mu_8\text{-Se})(\mu_5\text{-Se})_{12}(PEt_3)_{12}$

10

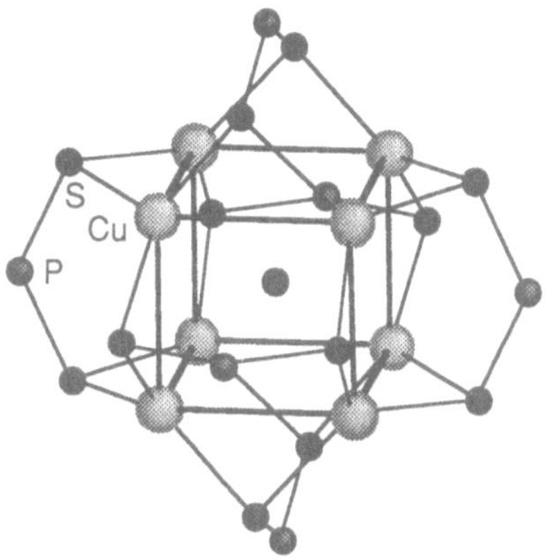

$Cu_8(\mu_8\text{-}S)(S_2P(OiPr_2))_6$

11

ally related 260-MVE compound $Pd_{20}As_{12}(PPh_3)_{12}$, mentioned previously, or the 282-MVE cluster $Ni_{21}Se_{12}(PPhEt_2)_{12}$ which contains a metal-centered Ni_9 cube [41]. A cubic $Cu_8(\mu_8\text{-}Se)$ unit is also present in the large cluster $Cu_{70}Se_{35}(PEt_3)_{22}$ [50].

The 134-MVE molecule **11** synthesized by Fackler and coworkers [51] has an idealized T_h point group symmetry with the eight Cu atoms located at the corners of a nearly perfect cube and the six di-isopropyl-dithiophosphate ligands bridging across each face of the cube. From symmetry arguments and MO calculations carried out on the non-centered 128-MVE copper Cu_8S_{12} species [35], Avdeef and Fackler have proposed that the AOs of the interstitial S atom interact with a_{1g} and t_{1u} MOs of the Cu_8 cage derived from combinations of the lowest energy empty AOs on each trigonally coordinated Cu atom. The long Cu-(μ_8-S) contacts measured in this species, 2.694 Å, suggest that the copper-sulfur bonding is somewhat ionic in character, as confirmed by Fenske-Hall MO calculations which indicate a highly negatively charged interstitial S atom [51]. The long Cu-Cu separations, 3.105 Å, are considerably longer than the Cu-Cu distances measured in the non-centered 128-MVE copper Cu_8S_{12} species (see above) [32–34]. Despite these long distances which might indicate no M-M bonding in this cluster, the Cu atoms do not obey the 16-electron rule, nor the 18-electron rule. Indeed, each copper center (formally Cu(I) d^{10}) has an average of 17 electrons. No Jahn-Teller distortion is observed however. A localized "frozen" picture would be to bind the encapsulated S atom tetrahedrally to four 18-electron metal atoms, leaving the four other metal atoms with 16 electrons.

Very recently, Henkel and coworkers reported the 128-MVE octanuclear heterometallic cluster $[Cu_4Mn_4(\mu_4\text{-}S)(\mu\text{-}Si\text{-}C_3H_7)_{12}]^{2-}$ also containing an interstitial sulfur atom [52]. The Cu_4Mn_4 cube is severely distorted in such a way that the encapsulated S^{2-} ion is tetrahedrally bound to the Mn atoms. Magnetic measurements indicated that the Mn atoms are strongly antiferromagnetically coupled [52], and an open-shell electron configuration is expected for this compound.

The same $M_8(\mu_4\text{-}E)$ arrangement is observed in the 144-MVE cadmium compounds characterized by Dance et al. such as $[Cd_8(\mu_4\text{-}S)(SePh)_{16}]^{2-}$ [53]. In these species, each cadmium center is tetrahedrally ligand-surrounded and follows the 18-electron rule. No M-M bonding is observed in agreement with their MVE count ($18 \times 8 = 144$).

8
Condensed Cubic Architectures

Further reduction of the ligand to metal ratio leads to a situation where not only ligands but also metal atoms have to be shared between clusters, creating either spherical (see for instance $Pd_{20}(\mu_5\text{-}As)_{12}(PPh_3)_{12}$ (**4**) mentioned earlier) or nonspherical clusters. Amid the latter, oligomers and polymers made of condensed octahedral metal units via corners, edges, or faces are well documented both in molecular and solid state chemistry [8, 54]. On the other hand, the one-dimensional growth of metal cubic clusters which occurs through the stacking of cubic arrays is very seldom observed, and is based on distorted rather than regular cubic units.

The compound $[Co_{14}Bi_8(CO)_{20}]^{2-}$ (**12**) recently reported by Whitmire and Eveland, constitutes one example of condensed cubic species. It can be viewed as a fusion of two of the tetragonal prismatic compounds $[Co_9Bi_4(CO)_{16}]^{2-}$ (**7**) which share a tetracobalt face [46].

Assuming that the closed-shell favored electron count for cluster **7** is 136 (see above), application of the principle of *polyhedral fusion* [55] yields an electron count of 208 MVEs ($136 \times 2 - 64 = 208$) or 210 MVEs ($136 \times 2 - 62 = 210$) for 12

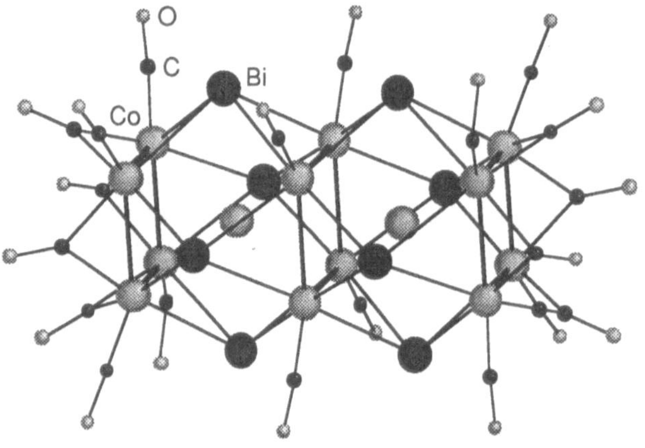

$[Co_{14}(\mu_5\text{-}Bi)_8(\mu\text{-}CO)_8(CO)_{12}]^{2-}$

12

(64 and 62 being the usual electron counts for square M_4 compounds [55, 56]). This is 16 and 18 electrons more than the observed count of 192 MVEs, respectively.

The bonding in this species was recently analyzed [47]. Interactions between the Co_i atoms and their $Co_{12}Bi_8$ host are comparable to those observed in its parent compound **7**. The high-lying diffuse 4s and 4p combinations of the two interstitial Co atoms interact strongly, whereas the low-lying d orbitals interact weakly with the top of the d-block of the $Co_{12}Bi_8$ host. A significant HOMO-LUMO gap is computed for the hypothetical count of 210 MVEs corresponding to a complete filling of this band. However, as quoted above for **7**, the depopulation of the upper MOs, which are Co_i-Co_s and Co_i-Bi antibonding, enhances the stability of the cluster by strengthening the bonding between the interstitial Co atoms and the $Co_{12}Bi_8$ fragment in the 192-MVE cluster **12**. EH and DFT results show a very small HOMO-LUMO gap (0.03 eV) for the observed electron count.

Another example based on the same condensation principle, namely the chinese lantern-shaped $Ni_{21}Se_{14}(PEt_2Ph)_{12}$ (**13**), has been synthesized by Fenske and Magull [57]. Its structure consists of a selenium-tetracapped metal-centered Ni_{13} cuboctahedron fused with two Ni_8 cubes along the four-fold axis. Assuming a full participation of the 3d Ni_i AOs to the bonding, the polyhedral fusion principle [55] leads to a favored closed-shell MVE count of 286 (170 + $120 \times 2 - 62 \times 2$), in relatively good agreement with the actual number of 290 MVEs counted for **13**. However, EH calculations suggest an electron-rich character with no HOMO-LUMO gap for such an electron count [47]. Here

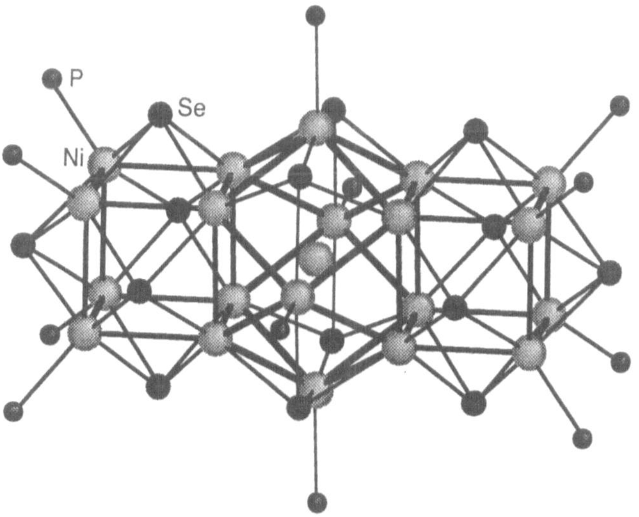

$Ni_{21}(\mu_4\text{-}Se)_{14}(PEt_2Ph)_{12}$

13

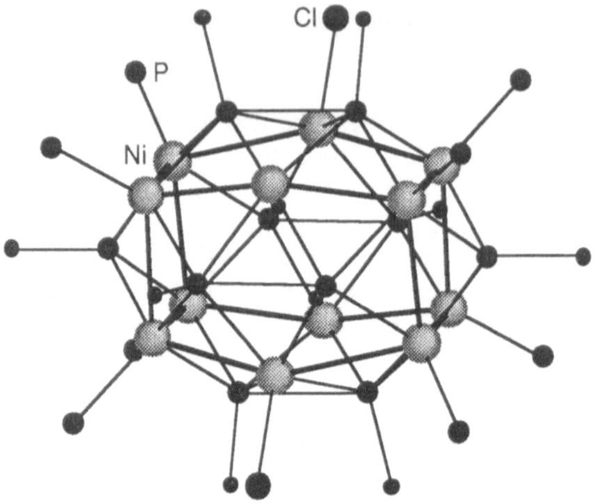

$Ni_{12}(\mu_4\text{-}PPh)_2(\mu_6\text{-}P_2Ph_2)_4(PHPh)_8(Cl)_2$

14

again, as noted for compounds **7** and **12**, the HOMO/LUMO region is made of MOs which are slightly antibonding between the Ni_i atom and its surrounding Ni_s atoms of the cuboctahedral part of the cluster.

The species $Ni_{12}(\mu_4\text{-}PPh)_2(\mu_6\text{-}P_2Ph_2)_4(PHPh)_8(Cl)_2$ (**14**) provides another relevant example for this category of large clusters [14]. Such a compound may be described as resulting from the condensation of two noncentered distorted cubic M_8 via a rather swelled metallic square. On the basis of the polyhedral fusion principle [55], cluster **14** should possess 184 ($128\times2-72$) MVEs. Counting the ($\mu_6\text{-}P_2Ph_2$) bridging ligand as 6-electron donor and the terminal PHPh ligand as 3-electron donor [58], it contains 178 MVEs, i.e., six fewer than expected.

Compounds **12**, **13**, and **14** are examples of condensed clusters suggesting the posibility of larger clusters with additional layers, leading ultimately to an infinite one-dimensional network. Our EH calculations carried out on hypothetical polymers and oligomers made of fused metal-centered tetragonal prismatic M_9E_4 units indicate that a full metallic situation (i.e., no significant HOMO-LUMO gap for any realistic electron count) is reached as soon as three such units are fused [47].

9 Metal Cubic Units in Solid State Chemistry

Cubic clustering is not limited to molecular species. Similar cubic architectures are also observed in solid state chemistry, such as in natural ($(Fe, Co, Ni)_9S_8$) [59], synthetic (Co_9S_8) (**15**) [60] pentlandites, and in djerfisherites such as $K_6LiFe_{24}S_{26}Cl$ [61] or $Ba_6Co_{25}S_{27}$ [62]. Metallic behavior is usually observed in these compounds presenting $M_8(\mu_4\text{-}S)_6$ entities, which exhibit a narrow range of electron counts.

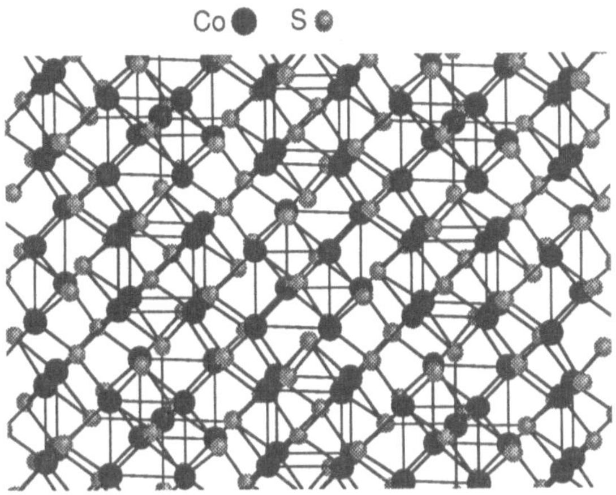

Co_9S_8

15

In Co_9S_8 for instance, distinguishable hexacapped cubic $Co_8(\mu_4\text{-}S)_6$ clusters are found surrounded by octahedrally coordinated CoS_6 units. The electronic properties of this material were examined in detail by Burdett and Miller some years ago [24]. With 107 MVEs, the $Co_8(\mu_4\text{-}S)_6$ units can be compared to the 108 and 109-MVE molecular analogs $[Co_8(\mu_4\text{-}S)_6(SPh)_8]^{4-,5-}$ [18]. Despite the fact that the Co-Co separations are slightly shorter in the former (by ca. 0.16 Å), the Co-Co and Co-S bondings are rather similar in both species [24].

Cubic 144-MVE $[Cu_8(\mu_4\text{-}Te_2)_6Te_8]^{4-}$ and 143-MVE $[Cu_8(\mu_4\text{-}Te_2)_6Te_8]^{3-}$ clusters of T_h symmetry centered by an alkali or alkali-earth atom can be "isolated" in the solid state materials $K_4Cu_8Te_{11}$, $A_2BaCu_8Te_{10}$ (A=K, Rb, Cs), $A_3Cu_8Te_{10}$ (A=Rb, Cs), and $AA'_2Cu_8Te_{10}$ (A, A'=K, Rb, Cs). In these compounds, recently reported by Kanatzidis et al. [63, 64], the Cu atoms adopt a distorted tetrahedral coordination geometry and follow the 18-electron rule for the count of 144 MVEs [64]. Ab initio Hartree-Fock calculations performed by Poblet et al. suggest that the chemical bonding and the topology in the cubic $Cu_8(Te_2)_6$ units is strongly different from that computed for the metallocarbohedranes $M_8(\mu_4\text{-}C_2)_6$ previously mentioned, due to the different electronic properties of the Te_2 and C_2 ligands [65].

10 Conclusion

In this review we have endeavored to outline the way in which the electronic structures of the transition metal compounds based on a cubic architecture may be rationalized, using results obtained from molecular orbital calculations. This has led to some interesting extensions of the electron-counting PSEP rules. The examination of the different cubic cluster categories has shown that the cluster topology is highly dependent on several parameters which can be calibrated.

The noncentered hexacapped $M_8(\mu_4\text{-E})_6$ architecture is allowed for a broad range of electron counts. The upper limit (120 MVEs) corresponds to a "magic" closed-shell electronic configuration secured by a significant HOMO-LUMO gap. Larger electron counts would destabilize and break the metal cubic core. On the other hand, lower electron counts are possible, down to to the hypothetical limit of 76 MVEs, particularly with π-donor terminal ligands.

With a metal atom or a main-group atom at the center of the metal cube, the $M_9(\mu_4\text{-E})_6$ and $M_8(\mu_8\text{-E}')(\mu_4\text{-E})_6$ cubic species have electron counts generally higher than those of the empty cubic related structures. A wide range of counts up to 130 MVEs has been reported so far depending on the magnitude of the interaction of the interstitial atom with its metallic cubic host and the nature of the capping E ligands (either bare or substituted). Among these electron counts, few, such as 124 with an interstitial metal atom, or 120 or 122 with an interstitial main group atom, correspond to closed-shell electronic configurations, but the majority of the observed counts correspond to open-shell electronic configurations.

Most of the "electron-deficient" centered or noncentered cubic clusters generally exhibit an open-shell electronic configuration, or at least a small HOMO-LUMO gap, which do not lead to a strong Jahn-Teller distortion. This is due to the high connectivity of the different atoms constituting the cluster core. Making new bonds induces a lengthening or a breaking of other bonds. This is reminiscent of structural changes encountered in solid state chemistry. Thus, it is interesting to mention that two situations can coexist from the point of view of electron-counting for these cubic species. The first situation is that generally observed for stable molecular systems, i.e., closed-shell electron configurations corresponding to "magic" MVE counts with a significant HOMO/LUMO gap. The second one, common in extended structures, allows a range of possible electron counts with no significant gap between the skeletal frontier orbitals and, consequently, open-shell electronic configurations are preferred. The high connectivity between the atoms forming the cluster core diminishes the Jahn-Teller instability. It is predictable that the occurrence of variable electron counts should become more frequent for the same architecture as the nuclearity of the cluster increases, because of decreases in the HOMO-LUMO separation and the electron pairing energy. These result from increasing delocalization and a progressive change from discrete to quasi-continuous energy levels. This is particularly true for the condensed cubic compounds.

Most of the research on transition metal cubic clusters has concentrated on their synthesis and their structural characterization. Little attention has been devoted so far to their reactivity and physical properties. The elucidation of their bonding may in future contribute to a deeper understanding of these aspects. The stability of the cubic architecture for a large gamut of electron counts is now well understood and largely confirmed experimentally. Fenske has shown for instance that some of the noncentered $M_8(\mu_4\text{-E})_6$ species can undergo redox processes without altering their metal cubic arrangement [8]. This ability to behave either as an electron sink or an electron reservoir could be exploited vis-a-vis donor or acceptor molecules for designing new materials with interesting properties. Though the octahedral fragment is the most abundant in cluster solid state chemistry [54], new conducting and/or magnetic extended structures resulting from the assemblage of cubic units can be envisaged. Stable cubic complexes with either complete or incomplete terminal ligand sheath may serve for exchanging labile terminal ligands during reactions. Cubic clusters can be characterized by an electronic structure which is either "molecular" or "metallic" in nature (HOMO-LUMO gap vs open-shell electronic configuration). Accurate measurements of their physical properties may provide some indication of the size of a particle where the typical properties of bulk metal begin to appear [66]. We hope this review will foster further studies on this kind of compound.

Acknowledgments. Thanks are expressed to our collaborators who participated in some of our works mentioned in this review: Dr. E. Furet, R. Gautier, Dr. A. Le Beuze, and B. Zouchoune. Pr. K.H. Whitmire and Pr. D. Fenske are acknowledged for providing results prior to publication. We also thank Pr. D.M.P. Mingos for having critically read the manuscript. Drawings of compounds were made based on X-ray crystallographic data using the Ca.R.Ine Cristallographie 3.0.1 program (C. Boudias and D. Monceau, 1989–1994).

11 References

1. Cotton FA (1966) Q Rev Chem Soc 20:389
2. McPartlin M, Eady CR, Johnson BFG, Lewis J (1976) J Chem Soc Chem Commun 883
3. Vaughan PA (1950) Proc Natl Acad Sci U S 36:461
4. The term PSEP was first introduced by Mason R, Thomas KM, Mingos DMP (1973) J Am Chem Soc 95:3802. For a complete description of the PSEP theory see for instance: a) Wade K (1976) Adv Inorg Chem Radiochem 18:1; b) Wade K (1981) In: Johnson BFG (ed) Transition metal clusters. Wiley, New York, p 193; c) Mingos DMP (1984) Acc Chem Res 7:311; d) Mingos DMP, Johnston RL (1987) Struct Bonding 68:29; e) Mingos DMP, Wales DJ (1990) Introduction to cluster chemistry. Prentice-Hall, Englewood Cliffs, New Jersey
5. a) Hoffmann R (1963) J Chem Phys 39:1397; b) Hoffmann R, Lipscomb WN (1962) J Chem Phys 36 21:79
6. See for example: a) Ziegler T (1991) Chem Rev 91:651; b) Rösch N, Pacchioni G (1994) In: Schmid G (ed) Metal clusters and colloids. Weinheim, p 5; c) Ziegler T (1995) Can J Chem 73:743; d) Kohn W, Becke AD, Parr RG (1996) J Phys Chem 100:12|974
7. See for example: a) Cotton FA, Haas TE (1964) Inorg Chem 3:10; b) Johnston RL, Mingos DMP (1985) Inorg Chem 25:1661; c) Mealli C, Lopez JA, Sun Y, Calhorda MJ (1993) Inorg Chim Acta 213:199; d) Lin Z, Williams ID (1996) Polyhedron 15:3277
8. Fenske D (1994) In: Schmid G (ed) Metal clusters and colloids. Weinheim, p 212
9. Lower LD, Dahl LF (1976) J Am Chem Soc 98:5046
10. Johnston RL, Mingos DMP (1985) J Organomet Chem 280:407
11. Fenske D, Basoglu R, Hachgenei J, Rogel F (1984) Angew Chem Int Ed Engl 23:160
12. Fenske D (unpublished results)
13. Fenske D, Magull J (1990) Z Naturforsch 45b:21
14. Fenske D, Krautscheid H, Müller M (1992) Angew Chem Int Ed Engl 31:321
15. Fenske D, Hachgenei J, Ohmer J (1985) Angew Chem Int Ed Engl 24:706
16. Fenske D, Ohmer J, Hachgenei J, Merzweiler K (1988) Angew Chem Int Ed Engl 27:1277
17. Saak W, Pohl S (1991) Angew Chem Int Ed Engl 30:881
18. Christou G, Hagen KS, Bashkin JK, Holm RH (1985) Inorg Chem 24:1010

19. Junghans C, Saak W, Pohl S (1985) J Chem Soc Chem Commun 2327
20. a) Pohl S, Opitz U (1993) Angew Chem Int Ed Engl 32:863; b) Pohl S, Barklage W, Saak W, Opitz U (1993) J Chem Soc Chem Commun 1251
21. Pohl S, Saak W (1984) Angew Chem Int Ed Engl 23:907
22. Fenske D, Hachgenei J, Rogel F (1984) Angew Chem Int Ed Engl 23:982
23. Arif AM, Jones RA, Heaton DE, Nunn CM, Scwab ST (1988) Inorg Chem 27:254
24. Burdett JK, Miller GJ (1987) J Am Chem Soc 109:4081
25. Hoffman GG, Bashkin JK, Karplus M (1990) J Am Chem Soc 112:8705
26. Rösch N, Ackermann L, Pacchioni G (1993) Inorg Chem 32:2963
27. Furet E, Le Beuze A, Halet J-F, Saillard J-Y (1994) J Am Chem Soc 116:274
28. See for example: Albright TA, Burdett JK, Whangbo MH (1985) In: Orbital interactions in chemistry. Wiley, New York, p 258
29. a) Hoffmann R (1982) Angew Chem Int Ed Engl 21:711; b) Evans DG, Mingos DMP (1985) J Organomet Chem 295:389
30. Fenske D, Fleischer H, Persau C (1989) Angew Chem Int Ed Engl 28:1665
31. a) Mingos DMP (1985) J Chem Soc Chem Commun 1353; b) Mingos DMP, Lin Z (1988) J Chem Soc Dalton Trans 1657
32. McCandlish LE, Bissell EC, Coucouvanis D, Fackler JP, Knox K (1968) J Am Chem Soc 90:7357
33. Hollander FJ, Coucouvanis D (1974) J Am Chem Soc 96:5646
34. Hollander FJ, Caffery ML, Coucouvanis D (1974) 167th National Meeting of the American Chemical Society Los Angeles Calif INOR 73
35. Avdeef A, Fackler JP (1978) Inorg Chem 17:2182
36. See for example: a) Guo BC, Kerns KP, Castleman AW Jr (1992) Science 255:1411; b) Guo BC, Wei S, Purnell J, Buzza S, Castleman A W Jr (1992) Science 256:515; c) Wei S, Guo BC, Purnell J, Buzza S, Castleman A W Jr (1992) Science 256:818; d) Wei S, Guo BC, Purnell J, Buzza S, Castleman A W Jr (1992) J Chem Phys 96:4166; e) Cartier SF, Chen ZY, Walder GJ, Sleppy CR, Castleman A W Jr (1993) Science 260:195; f) Pilgrim JS, Duncan MA (1993) J Am Chem Soc 115:6958; g) Cartier SF, May BD, Castleman AW Jr (1994) J Chem Phys 100:5384; h) Cartier SF, May BD, Castleman AW Jr (1994) J Am Chem Soc 116:5295
37. Bénard M, Rohmer M-M, Poblet J-M, Bo C (1996) J Phys Chem 99:16913 and references cited therein
38. Zebrowski JP, Hayashi RK, Bjarnason A, Dahl LF (1992) J Am Chem Soc 114:3121
39. Brennan JG, Siegrist T, Stuczynski SM, Steigerwald ML (1989) J Am Chem Soc 111:9240
40. Fenske D, Persau C (1991) Z Anorg Allg Chem 593:61
41. a) Fenske D, Vogt K (unpublished results); b) Vogt K (1994) PhD Dissertation University of Karlsruhe Germany
42. Fenske D, Ohmer J, Merzweiler K (1988) Angew Chem Int Ed Engl 27:1512
43. Wheeler RA (1990) J Am Chem Soc 112:8737; (1991) J Am Chem Soc 113:4046
44. Nomikou Z, Schubert B, Hoffmann R, Steigerwald ML (1992) Inorg Chem 31:2201
45. Furet E, Le Beuze A, Halet J-F, Saillard J-Y (1995) J Am Chem Soc 117:4936
46. Whitmire KH, Eveland JR (1994) J Chem Soc Chem Commun 1335
47. Zouchoune B, Ogliaro F, Halet J-F, Saillard J-Y, Eveland JR, Whitmire KH Inorg Chem (submitted for publication)
48. Johnston RL, Mingos DMP (1985) J Organomet Chem 280:419
49. Gautier R, Halet J-F, Saillard J-Y (unpublished results)
50. Fenske D, Krautscheid H (1990) Angew Chem Int Ed Engl 29:1452
51. Liu CW, Stubbs T, Staples RJ, Fackler JP Jr (1995) J Am Chem Soc 117:9778
52. Stephan H-O, Kanatzidis MG, Henkel G (1996) Angew Chem Int Ed Egl 35:2135
53. Lee GSH, Fisher KJ Craig DC, Scudder ML, Dance IG (1990) J Am Chem Soc 112:6435
54. Simon A (1994) In: Schmid G (ed) Metal clusters and colloids. Weinheim, p 373
55. Mingos DMP (1983) J Chem Soc Chem Commun 706
56. a) Halet J-F, Hoffmann R, Saillard J-Y (1985) Inorg Chem 25:1695; b) Halet J-F (1995) Coord Chem Rev 143:637
57. Fenske D, Magull J (1991) Z Anorg All Chem 594:29
58. Bohle DS, Jones TC, Rickard CEF, Roper WR (1986) Organometallics 5:1612

59. a) Vaughan DJ, Craig JR (1978) Mineral chemistry of metal sulfides. Cambridge University Press, New York; b) Rajamani V, Prewitt CT (1973) Can Mineral 12:178; c) Rajamani V, Prewitt CT (1975) Am Mineral 60:39
60. a) Rajamani V, Prewitt CT (1975) Can Mineral 13:75; b) Kim K, Dwight K, Wold A, Chianelli RR (1981) Mater Res Bull 16:1319; c) Pasquariello DM, Kershaw R, Passaretti JD, Dwight K, Wold A (1984) Inorg Chem 23:872
61. a) Tani BS (1977) Am Mineral 62:819; b) Czamanske GK, Erd RC (1979) Am Mineral 64: 776
62. Snyder GJ, Badding ME, DiSalvo FJ (1992) Inorg Chem 31:2107
63. Park Y, Kanatzidis MG Chem Mater (1991) 3:781
64. Zhang X, Park Y, Hogan T, Schindler J, Kannewurf CR, Seong S, Albright TA, Kanatzidis MG (1995) J Am Chem Soc 117:10300
65. Poblet J-M, Rohmer M-M, Bénard M (1996) Inorg Chem 35:4073
66. For recent reviews see for instance: a) [54]; b) Corbett JD (1996) J Chem Soc Dalton Trans 575
67. Schmid G (1994) In: Schmid G (ed) Metal clusters and colloids. Weinheim, p 178

Metallaboranes

Thomas P. Fehlner

Department of Chemistry and Biochemistry, University of Notre Dame, Notre Dame, IN 46556, USA. *E-mail: Thomas.P.Fehlner.1@nd.edu*

Metallaboranes are true hybrid cluster systems that empirically join the polyhedral boranes with transition metal clusters. Although metallaborane chemistry, and metallacarborane chemistry, initially developed as a variation of metal-ligand coordination chemistry analogous to organometallic chemistry, it was the cluster electron counting rules and the associated isolobal principle that revealed interconnections between ostensibly unrelated molecules and defined the scope of the area. The subsequent synthetic developments, both rationalized and stimulated by these ideas, provide a convincing verification of the intrinsic chemical truths expressed by these rules. The variability in the isolobal behavior of a given metal fragment makes the chemistry associated with metallaboranes considerably richer than that of pure boranes or pure metal clusters of the same dimension. Consequently, metallaboranes require careful analysis of the connection between geometry and electronic structure. In this regard, the emerging metallaborane chemistry of the early transition metals provides interesting variations on the theme of cluster electron counting.

Keywords: metallaboranes, boranes, carboranes, clusters

1 Introduction

1.1 Metallaboranes – A Definition

Metallaboranes may be defined as compounds that contain direct metal-boron bonding. By their nature they are hybrid compounds not unlike organometallic compounds. That being so, this author suggested they be described by the adjective "inorganometallic" in order to connect them formally to compounds with direct transition metal-main group element (other than carbon) bonding and to relate their chemistry to organometallic chemistry [1]. Thus, metallaborane chemistry is also a part of inorganometallic chemistry. Ambiguities exist in such a division but then organoboron chemistry itself is considered part of organometallic chemistry by some authors [2].

The emphasis of this chapter is on metallaboranes but metallacarboranes will be included as a subset as they also contain metal-boron bonds. It must be noted that many of the initial advances involved carboranes rather than boranes and there is no intent to slight an important and large set of compounds. But the purpose of this chapter is to analyze the impact of the electron counting rules on a compound type rather than the exhaustive compilation of compounds and their properties. Both metallaborane chemistry and metallacarborane chemistry have been comprehensively reviewed previously [3–10] and this chapter presents a somewhat personal evaluation of metallaboranes in light of the electron counting rules for clusters.

Independently of the descriptors used, metallaborane compounds constitute an experimental link between transition metal clusters, borane cages, organometallic compounds and solid state metal borides [11]. The existence of these compounds provides ample justification for both the cluster electron counting rules as well as the isolobal principle [12–16]. As suggested by the following, it is doubtful that this subset of cluster chemistry would have developed as rapidly as it has without the guidance of these ideas. The rules provide a foundation to our understanding of the area but they do not limit the chemistry. Indeed, there is metallaborane chemistry falling outside their purview that is only appreciated with the perspective provided by the electron counting rules.

1.2 Chapter Outlook

The chapter is organized sequentially around three perspectives: that of a classical coordination chemist, that of an organometallic chemist, and that of a cluster chemist. The idea is to give the reader some sense of the historical development of the area and the evolving understanding of these compounds. In the process, the unifying effects of the structural and electronic principles expressed in the cluster electron counting rules and the isolobal analogy are highlighted, particularly for compounds that can only reasonably be considered as clusters.

In the case of cluster compounds, it is important that the reader appreciates how a sense of confident interpretation of molecular formula in terms of cluster structure both simplified and stimulated the synthetic chemistry. Spectroscopic characterization in the light of the cluster electron counting rules often gives a useful sense of structure. It is not that spectroscopic characterization gives unambiguous structural information, as it often does not. However, the exceptions to the rules can usually be recognized as such even if knowledge of the actual structure requires the definition of X-ray diffraction experiment. A limited set of possible structures of the same genre fosters the chemical rationalizations, often not published, that move the chemistry forward. Just as the working organic chemist pushes electrons seeking rationalizations and extending reaction pathways, so the cluster chemist assembles deltahedra consonant with the available cluster bonding electrons in order to rationalize structure and extend chemistry.

Finally some of the author's recent chemistry serves as a foil to a look beyond the cluster bonding rules into an area of cluster chemistry only now emerging. Thus, the electron counting rules are required to define a delocalized unsaturated metallaborane even though they do not predict or require the existence of such a bonding system.

2 Boranes as Ligands

2.1 Metal-Ligand Chemistry

The systematic development of transition metal chemistry began with the coordination chemistry of positively charged metal ions. In coordination compounds the metal center acts as a classic Lewis acid with properties empirically described by the Irving-Williams series and similar approaches. The ligand, as the complementary Lewis base, adds in a stoichiometric ratio controlled by the electronic requirements of the metal, the steric requirements of the ligand, the "ligand field" generated by the set of ligands, and other factors. Naturally many ligands possess atoms found to the right of carbon in the periodic table as lone pairs constitute intrinsic sites of Lewis basicity.

Both to attack structural problems and to investigate the spectroscopic properties associated with the metal center, an intense effort was put into the development of various ligand types. Control of ligand properties allowed control of metal chemistry and, in turn, variation in metal properties permitted the chemistry of coordinated ligands to be varied in systematic and useful fashions. Valuable extensions into synthesis, catalysis, and biomimetic modeling have followed. Indeed, the development of new ligands is far from over as is illustrated by recent forays into ligand arrays and dendridic complexes [17].

Like transition metals, atoms to the left of carbon possess fewer electrons than valence orbitals. The chemistry, if not dominated by interactions with Lewis bases, is certainly strongly associated with the Lewis acidity of the compounds. Hence, one does not automatically include ligating ability with the attributes of boranes, for example. The fact that borane anions can act as ligands in the same sense as halogen anions was recognized relatively early and a brief account of some representative chemistry serves to introduce the first type of metallaborane – a simple metal-ligand coordination compound.

2.2
Coordination Chemistry of Borane Anions

The simplest borane anion is $[BH_4]^-$ which is isoelectronic with CH_4. The coordination chemistry of the latter is of intense interest in terms of activating the CH bond but requires a metal center with a very specialized binding site [18–23]. The former, with its net negative charge and polarized BH bonds, has an extensive coordination chemistry. Thus, in addition to metal salts, a significant number of complexes containing covalently bound $[BH_4]^-$ ligands have been known since the 1950s [24] with the first reference to $Al(BH_4)_3$ appearing in 1939. Since that time complexes of a great number of the transition metals, lanthanides, and actinides have been described [25]

As it is isoelectronic with $[F]^-$, $[BH_4]^-$ can be viewed as a simple replacement for a halogen in a transition metal complex. In fact halide displacement by borohydride ion constitutes a preparative route to these compounds. However, the hydrogens of $[BH_4]^-$ introduce an element of structure not present in the halide complex in that one, two or three hydrogens are found to bridge between the boron and metal atoms in the static structure. In a typical borohydride complex such as $[(CO)_4Mo(BH_4)]^-$ (Fig. 1a), there are two bridging B-H-M interactions and the $[BH_4]^-$ functions as a 4-electron donor to the metal center (al-

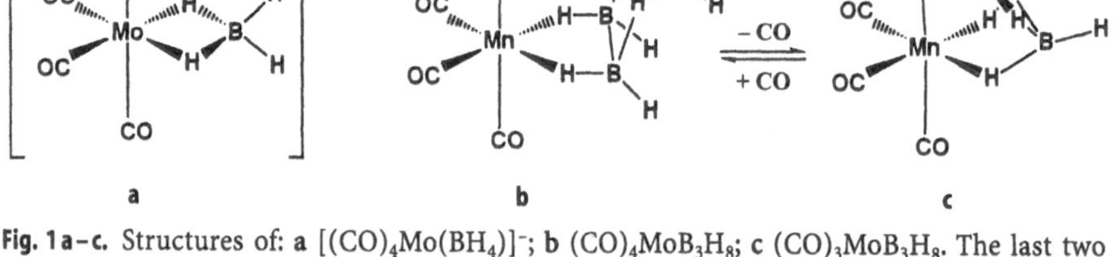

Fig. 1a–c. Structures of: **a** $[(CO)_4Mo(BH_4)]^-$; **b** $(CO)_4MoB_3H_8$; **c** $(CO)_3MoB_3H_8$. The last two compounds are connected via an equilibrium involving the loss and addition of a CO ligand to the Mn atom

though the compounds chosen for examples follow the 18-electron rule in so far as the metal center is concerned, complexes which are paramagnetic are known as well). For almost all mononuclear borohydride complexes, the proton NMR exhibits a single averaged resonance for the BH protons even at low temperature suggesting that the M-H component of the B-H-M bridge is weaker than the B-H component and that the energy barrier associated with breaking the M-H bond is small. This behavior is consistent with viewing the $[BH_4]^-$ moiety as a ligand with its own distinct identity in both the bound and unbound states.

The difference between $[BH_4]^-$ and $[F]^-$ is that four protons found in the nucleus of the latter are disposed in a tetrahedral array around the central boron nucleus. The "protonated lone pairs" of $[BH_4]^-$ are much more focused than the lone pairs of the halide albeit of reduced basicity. This crude picture suggests a B-H bond as a potential 2-electron donor to a vacant metal orbital thereby associating Lewis base character with the B-H bonds of boranes. Thus, other borane anions or even boranes should coordinate via B-H bonds as Lewis bases to Lewis acidic transition metal sites.

A substantial class of compounds of this type are derived from the $[B_3H_8]^-$ ion and one fascinating example serves to display the characteristics of these species [26, 27]. As shown in Fig. 1b, two B-H bonds of the $[B_3H_8]^-$ ligand in $(CO)_4Mn(B_3H_8)$ occupy two octahedral coordination sites of the 14-electron $[(CO)_4Mn]^+$ ion fragment just as the borohydride ligand does in $[(CO)_4Mo(BH_4)]^-$. Hence, the $[B_3H_8]^-$ ligand can likewise be viewed as a 4-electron donor to the metal but, as the two B-H bonds utilized are on distinct boron atoms, the ligand behaves like a chelating diphosphine.

On heating $(CO)_4Mn(B_3H_8)$ a CO ligand is lost to yield $(CO)_3Mn(B_3H_8)$ (Fig. 1c) in a reaction that is reversible. This presents an unambiguous example of the displacement of a Lewis base by a B-H bond and confirms the intrinsically basic character of this fundamental component of any borane. Furthermore, it defines the protonated lone pair of the B-H bond as a 2-electron donor as is a conventional lone pair.

There are even more spectacular examples of borane and carborane anions utilizing B-H bonds as donor sites in, for example, $(CO)_3MnB_8H_{13}$ where an open $[B_8H_{13}]^-$ cage uses three exo-BH bonds to chelate the $[(CO)_3Mn]^+$ in exactly the same manner as the $[B_3H_8]^-$ fragment does in $(CO)_3Mn(B_3H_8)$ [28]. A similar situation is found in the so-called exo-nido complex involving a carborane ligand bound to a Rh center via the external cage B-H terminal bonds [29–32]. Recently, η^5-carborane ligands (see below) have been found to incorporate such bonding modes in coordinating a variety of metal species [33]. Although many of these compounds can be viewed as clusters as well, e.g., $(CO)_4Mn(B_3H_8)$ can be viewed as a metallatetraborane analog of B_4H_{10}, the metal-ligand model is perfectly adequate for describing both formation and bonding.

2.3
Coordination Chemistry of Neutral Borane Base Adducts

Of course with the anionic boranes, one might argue that the dominant factor in the formation of the metal complexes is ionic, i.e., the Lewis base character

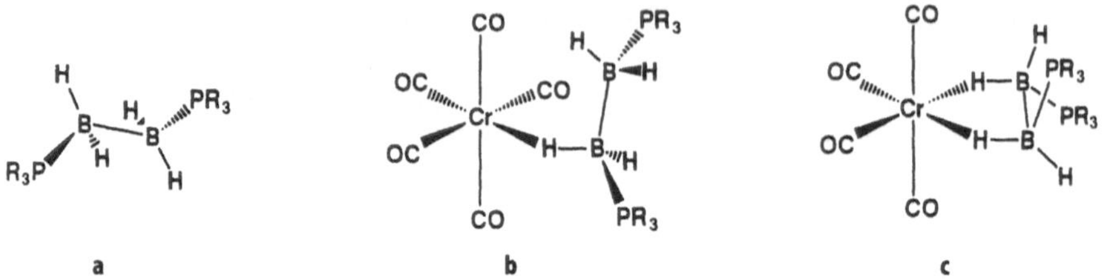

Fig. 2a–c. Structures of: **a** the $B_2H_4(PMe_3)_2$ ligand; **b** $(CO)_5Cr[B_2H_4(PMe_3)_2]$; **c** $(CO)_4Cr[B_2H_4(PMe_3)_2]$

of the B-H bond is a trivial component in the overall metal-borane binding. The interconversion of $(CO)_4Mn(B_3H_8)$ and $(CO)_3Mn(B_3H_8)$, discussed above, suggests that this may not be the case. More recently it has been demonstrated that a polar B-H bond in a neutral borane can be used directly as an effective donor to a metal thereby opening a potentially large area of new borane coordination chemistry. The bisphosphine diborane(4) neutral molecule (Fig. 2a) with an ethane-like geometry behaves as a unidentate ligand in $(CO)_5Cr[B_2H_4(PMe_3)_2]$ (Fig. 2b) and a bidentate ligand in $(CO)_4Cr[B_2H_4(PMe_3)_2]$ (Fig. 2c), which should be compared with the structure of $[(CO)_4Mo(BH_4)]^-$ (Fig. 1a) [34–36]. The five-membered CrHBBH ring of $(CO)_4Cr[B_2H_4(PMe_3)_2]$ is reminiscent of the five-membered ring formed with a chelating bidentate phosphine such as $R_2PCH_2CH_2PR_2$. No doubt additional stability is derived from ring formation in the case of the borane as well.

The literature contains other examples of compounds in which a borane can be viewed as a complex chelating ligand albeit their origins are not as straightforward. In fact, the first characterized example of a multinuclear metallaborane, $(CO)_{10}HMn_3B_2H_6$ (Fig. 3a), has a B_2H_6 ligand in an ethane-like geometry. Effectively the borane is $[B_2H_6]^{2-}$, which of course is $B_2H_4L_2$ with $L=[H]^-$. New, fascinating examples continue to be synthesized. For example, the novel, and extremely sensitive, dimeric compound $[(MeZn)_2B_3H_7]_2$ has recently been reported (Fig. 3b) [37]. Curiously, there are two distinct Zn environments but the two B_3H_7 fragments bridge (rather than chelate) one pair of MeZn frag-

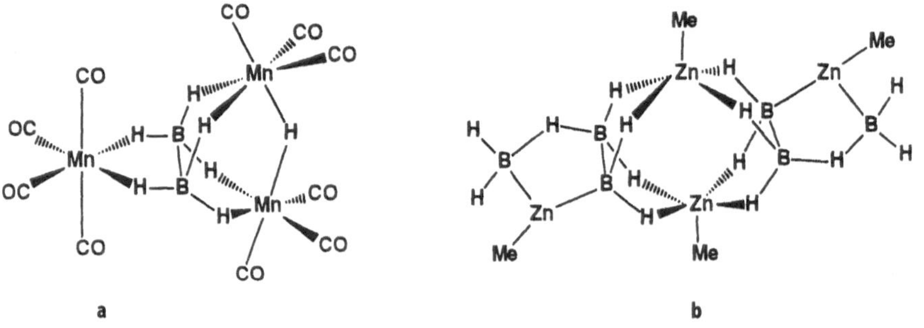

Fig. 3.a, d Structures of: **a** $(CO)_{10}HMn_3B_2H_6$; **b** $[(MeZn)_2B_3H_7]_2$

ments using four B-H bonds in the manner discussed. As weakly coordinating neutral ligands are of significant interest in terms of making "lightly stabilized" coordination compounds and clusters, interest in the utilization of various boranes as weakly bound ligands is expected to continue [38].

3
Boranes as Hydrocarbyl Ligand Mimics

3.1
Organometallic and Inorganometallic Chemistry

In an early review of metallaborane chemistry, almost all the compounds then known were viewed as metal ligand complexes [39]. At the same time, the connection with organometallic chemistry was recognized and employed in the organization of the structural chemistry in the manner of organic "ligands." For example, the binding of $[BH_4]^-$ to metal centers has been compared to that of a η^3-allyl ligand [25]. Most of the compounds discussed in the previous section could well be called organometallic analogs as the metals are in low oxidation states and the 18-electron rule rules. However, other aspects of metallaborane chemistry more closely mimic organometallic chemistry and constitute a part of main group-transition element chemistry that gives meaning to the term "Inorganometallic Chemistry", i.e., metallaborane chemistry can be considered as a variation of organometallic chemistry utilizing its electron counting formalisms.

3.2
Cyclopentadienyl Ligand Mimics

The distinctive character of organometallic chemistry originates in large part from the identification and utilization of organic ligands that utilize π bonding systems in coordinating to metal centers in η^n-modes. Likewise metallaborane chemistry, more precisely metallacarborane chemistry, received a quantum boost when Hawthorne recognized the operational similarities between the open five-membered face of the $[C_2B_9H_{11}]^{2-}$ ion and the cyclopentadienyl ion ($Cp^- = [\eta^5\text{-}C_5H_5]^-$) relative to its capabilities as a 6-electron π-ligand (Fig. 4a) [9, 10, 40]. A closely related area involves boron heterocycles, synthesized directly, e.g., 1,3-diborolenes [41, 42] or indirectly by "decapping" a polyhedral metallacarborane to yield the bound $[C_2B_3H_7]^{2-}$ planar ligand (Fig. 4b) [43], as Cp^- mimics. This chemistry has been reviewed extensively, indeed some is standard textbook material, and the following only touches on a couple of points of particular relevance to the chapter theme.

To be sure there are clear similarities between, e.g., $[C_2B_9H_{11}]^{2-}$ and $[C_5H_5]^-$, but the former is not simply an interchangeable replacement for the latter. The chemistry is related to metallocene chemistry but with its own distinct character. For example, the analog of ferrocene, $[Fe(C_2B_9H_{11})_2]^{2-}$ is known but it exists preferentially as $[Fe(C_2B_9H_{11})_2]^-$ in the Fe(III) oxidation state rather than in the Fe(II) state as found for ferrocene. One other different aspect concerns the

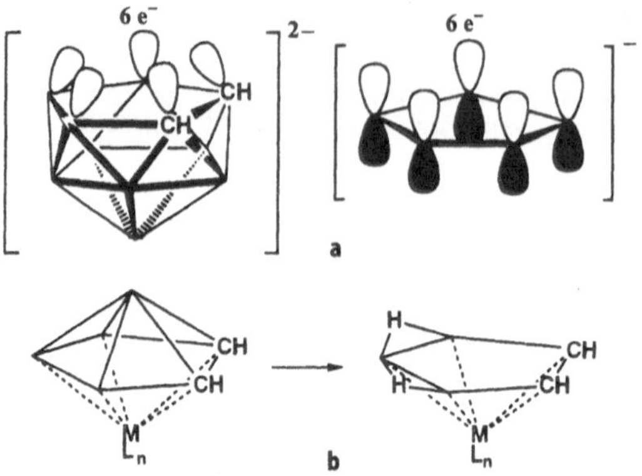

Fig. 4. **a** Comparison of the open face of the $[1,2\text{-}C_2B_9H_{11}]^{2-}$ ion and that of $[C_5H_5]^-$. **b** Formation of the planar $[C_2B_3H_7]^{2-}$ analog of π-bound $[C_5H_5]^-$ by the "decapitation" of a $C_2B_4H_6$ metallacarborane. Note that in the cluster frameworks the CH but not the BH vertices are labeled

greater variation in metal-carborane binding which is illustrated by a nickel carborane sandwich complex. Three compounds are known: $[Ni(1,2\text{-}C_2B_9H_{11})_2]^{2-}$, $[Ni(1,2\text{-}C_2B_9H_{11})_2]^-$, and $[Ni(1,2\text{-}C_2B_9H_{11})_2]$ with Ni oxidation states of II, III, and IV, respectively. Although each has the basic sandwich structure, they have distinctly different structures [44]. The first has a "slipped" structure, the second a symmetrical structure with the carbon atoms of the two cages transoid to each other, and the third a less symmetrical structure with the carbon atoms cisoid to each other (Fig. 5a). Furthermore, on heating $[Ni(1,2\text{-}C_2B_9H_{11})_2]$ the carbon atoms in the cage migrate to positions of higher stability thereby yielding more isomeric forms [45]. These are generally rather robust compounds and the scope of carborane ligand chemistry is very large indeed.

A greater metal binding capability is particularly evident in the chemistry of the planar $[C_2B_3H_5]^{4-}$ ligand. This species is not found in the free state but can be effectively generated in the form of $[L_nM(C_2B_3H_5)]^{2-}$ building blocks derived from $L_nM(C_2B_3H_7)$ (Fig. 4b). These units, with substituents other than hydrogen, have been utilized to make a variety of multidecker sandwich complexes in a systematic fashion. A spectacular example is the recently characterized hexadecker complex $[Cp^*Co(Et_2C_2B_3H_2Me)Co(Et_2C_2B_3H_3)]_2H_2Co$, $Cp^* = \eta^5\text{-}C_5Me_5$ (Fig. 5b) [46].

These distinctive variations on the theme of organometallic metallocene chemistry add valuable new dimensions to the chemistry of inorganometallic analogs of the ubiquitous Cp ligand. In fact, organometallic chemists now include carborane analogs of the Cp ligand in their synthetic repertoire [47, 48]. For completeness, let the reader recall that many other boron-containing rings as well as boranes can be viewed as analogs of various organometallic π complexes [49]. Even the boron analog of the classic π complex of ethylene is known [50, 51].

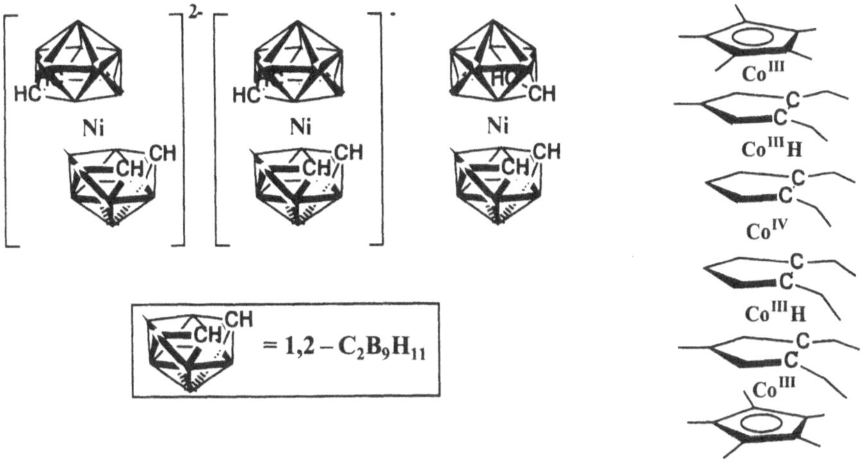

a b

Fig. 5. **a** Comparison of the structures of $[Ni(1,2\text{-}C_2B_9H_{11})_2]^{2-}$, $[Ni(1,2\text{-}C_2B_9H_{11})_2]^{-}$, and $[Ni(1,2\text{-}C_2B_9H_{11})_2]$. **b** The structure of the hexadecker sandwich complex, $[Cp^*Co(Et_2C_2B_3H_2Me)Co(Et_2C_2B_3H_3)]_2H_2Co$. Note that the vertices occupied by BH or B have not been labeled and that the hexadecker possesses Cp* ligands at both ends

3.3 Isoelectronic Borane and Hydrocarbyl Complexes

Although the substitution of a $[BH]^-$ fragment for a CH fragment in an organometallic molecule is by no means a small perturbation on the electronic structure and compound properties, the pairs of isoelectronic compounds generated by such a formal substitution present an excellent way in which to explore both species systematically. In fact, if the total number of electrons in a pair of compounds are the same, then the perturbation investigated is actually in the nuclear framework in terms of the spatial distribution of protons. For small molecules the results of proton shifts are of pedagogical interest. But even for larger systems the qualitative aspects of such perturbation in terms of nuclear shielding, for example, are straightforward [52]. Thus, we and others have used this approach to interrelate selected borane and hydrocarbyl complexes [53–56]. The following two brief examples makes the connection between metallaboranes and organometallic complexes explicit. Both compound types follow the 18-electron rule but show distinctive differences in structures and properties.

Note first, however, that the term “isoelectronic” is often used loosely to describe non-isoelectronic compounds in which the bonding features of interest are related in the sense of requiring the same number of electrons. Such use is fraught with difficulties and often gives rise to misconceptions [52].

The first example of an isoelectronic cluster comparison concerns the dependence of bridging hydrogen atom position on the identity of the main group atom, i.e., boron vs carbon, in the compounds of compositions

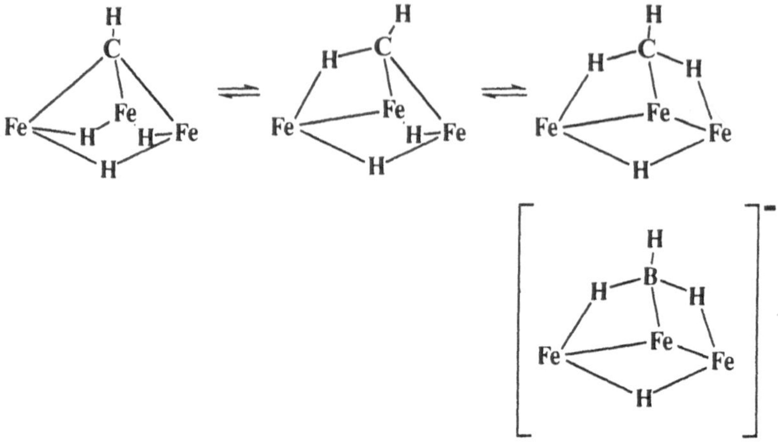

Fig. 6. The three tautomeric forms of $CH_4Fe_3(CO)_9$ and the single form of $[BH_4Fe_3(CO)_9]^-$. The terminal CO ligands are omitted for clarity

$CH_4Fe_3(CO)_9$ and $[BH_4Fe_3(CO)_9]^-$ [57–61]. The only difference between the clusters is the absence of a proton in the ferraborane relative to the hydrocarbyl complex. Although both cluster frameworks possess the expected main group atom capped metal triangle structure, the distribution of the non-terminal hydrogen atoms is different.

The compound $CH_4Fe_3(CO)_9$ exists in solution as an equilibrium mixture of three tautomers: $H_3Fe_3(CO)_9(\mu_3\text{-}CH)$, $H_2Fe_3(CO)_9(\mu_3\text{-}HCH)$, and $HFe_3(CO)_9(\mu_3\text{-}H_2CH)$ in the ratio of 86:13:1. The order of stability was unambiguously demonstrated by deprotonating the mixture to yield a single anion, $[HFe_3(CO)_9(\mu_3\text{-}HCH)]^-$. This anion, when protonated at low temperature, yields exclusively $HFe_3(CO)_9(\mu_3\text{-}H_2CH)$ as the sole kinetic product. Warming to room temperature results in conversion to the more stable tautomers shown in Fig. 6. These observations imply that protonation takes place preferentially at an exposed C-Fe edge of the cluster even though the Fe-Fe edge is the more basic site. In the case of the isoelectronic ferraborane, $BH_5Fe_3(CO)_9$, only one form, albeit fluxional, is observed [61]. This is $HFe_3(CO)_9(\mu_3\text{-}H_3BH)$. Deprotonation of the ferraborane leads to $[HFe_3(CO)_9(\mu_3\text{-}H_2BH)]^-$ which has the same hydrogen atom distribution as the least stable tautomer of $CH_4Fe_3(CO)_9$ (Fig. 6).

The difference between these isoelectronic species shows that, even though the geometric structures of the cores are the same, the C to B^- change affects the disposition of the edge-bridging hydrogens (and CO ligand orientations) on the M_3E tetrahedral framework. The simplest explanation of the observations is found in the distribution of electronic charge between main group and metal atoms which is given by the difference in the effective electronegativities of the atoms [62]. The bridging hydrogen atom is thus viewed as a proton seeking maximum electronic charge and the differing placements of the bridging atoms on the two tetrahedral frameworks reflect the differing skele-

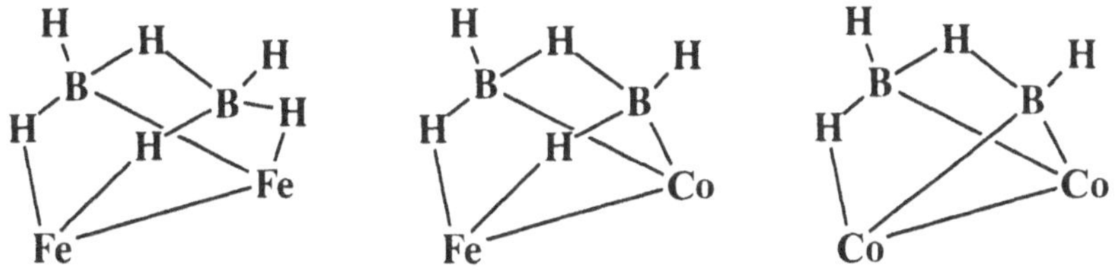

Fig. 7. Structures of $B_2H_6Fe_2(CO)_6$, $B_2H_5FeCo(CO)_6$, and $B_2H_4Co(CO)_6$. The three terminal CO ligands on each metal are not shown

tal charge distributions caused by the difference in the electronegativities of C and B. It follows that changing the transition metals should have like effects on the distribution of bridging hydrogens and, indeed, this is found to be the case [63].

The second example concerns the dependence of the Brönsted acidity of a bridging hydrogen on the identity of the transition metal atom, e.g., iron vs cobalt. The protonic character of a bridging hydrogen atom is well known [64, 65]. For the boranes, the acidity order as a function of cage size is known and includes the role of substituents [64]. For metal hydrides it is known that the acidity of the hydrogen atom in a mononuclear compound increases as one goes, e.g., from $HMn(CO)_5$ to $H_2Fe(CO)_4$ to $HCo(CO)_4$. These differences in acidity are large. For different ancillary ligands on the metal absolute acidities are substantially changed but the trend with metal identity remains the same [66–69]. We have investigated the situation for metallaboranes.

The isoelectronic compounds investigated are shown in Fig. 7 [70–72]. In the event only two could be used for the measurement of relative acidities as it was found that the $[B_2H_3Co_2(CO)_6]^-$ anion was too unstable to employ as a reactant. The competition reaction was investigated in both directions. The reaction at Eq. (2) occurred readily but that at Eq. (1) did not.

$$[B_2H_5Fe_2(CO)_6]^- + B_2H_5FeCo(CO)_6 \rightarrow B_2H_6Fe_2(CO)_6 + [B_2H_4FeCo(CO)_6]^- \quad (1)$$

$$B_2H_6Fe_2(CO)_6 + [B_2H_4FeCo(CO)_6]^- \rightarrow [B_2H_5Fe_2(CO)_6]^- + B_2H_5FeCo(CO)_6 \quad (2)$$

Thus, contrary to expectations based on acidity trends in mononuclear metal hydrides, $B_2H_6Fe_2(CO)_6$ is a stronger Brönsted acid than $B_2H_5FeCo(CO)_6$. One is forced to associate a greater acidity with the larger number of bridging hydrogen atoms in $B_2H_6Fe_2(CO)_6$. This is reasonable. In replacing FeH with Co one is effectively taking an unshielded proton and placing it in a metal nucleus where it is highly shielded by the metal core electrons. The effect of the proton in question on the valence properties is clearly going to be larger in the first case. Thus, if the process by which a proton is lost is strongly affected by the other hydrogens present, the acidity will be changed. As the extra hydrogen on $B_2H_6Fe_2(CO)_6$ is a bridging hydrogen, it probably perturbs the cluster bonding network and acts to stabilize the anion formed on proton loss.

4
Boranes as Metal Fragment Replacements

4.1
Borane and Metal Cluster Connections

By now it should be clear that the genesis of metallaborane chemistry is intimately linked with the developed ideas of coordination chemistry and its variant, organometallic chemistry. It was only in the early 1970s that a new view of a metallaboranes arose from a consideration of the metal atom as an intimate part of the cluster skeleton. That is, the borane, or carborane, was not viewed as a ligand bound to a metal center but rather the metal, or metals, was viewed as occupying a vertex of a larger heteroborane skeleton. From this new perspective, nearly all the compounds discussed above can be considered as heteroborane clusters.

Such a reformulation of metallaboranes would be an academic exercise, i.e., useless except for constructing challenging examinations, but for the fact that there are metallaboranes that cannot logically be considered as metal-ligand complexes. The view of metallaboranes as heteroborane clusters both permits unusual compounds to be easily rationalized and suggests new synthetic targets.

It all began with the nearly concurrent presentations of the cluster structure-cluster electron counting relationships [12, 65, 73–76] and the idea of isolobal transition metal and main group atom fragments [77]. These two powerful ideas gave rise to a new level of understanding of clusters of complex stoichiometry and structure. Disparate clusters such as $[B_6H_6]^{2-}$ and $[Ru_6(CO)_{18}]^{2-}$ were seen to be analogous in so far as cluster core bonding is concerned. Both have octahedral core geometries and seven cluster bonding pairs. The BH and $Ru(CO)_3$ fragments were recognized as isolobal in the sense that both, apparently, contribute three frontier orbitals of similar symmetry containing two electrons to cluster bonding. This implied the existence of mixed clusters of the type $[(BH)_n\{Ru(CO)_3\}_{6-n}]^{2-}$, n = 1–5, i.e., metallaborane clusters with one to five metal atoms related to the two parent species by the electron counting rules and the isolobal analogy. Thus, a synthetic challenge was generated. Would $[(BH)_3\{Ru(CO)_3\}_3]^{2-}$, for example, be a close relative of the parent borane and metal clusters or not? Just as important, these metallaboranes, if synthesized, would provide a set of compounds in which cluster properties systematically vary from those related to $[B_6H_6]^{2-}$ to those related to $[Ru_6(CO)_{18}]^{2-}$. Limited sets of these kinds of metallaboranes have been characterized and some are explored in different ways below.

4.2
Coordination Compound vs Cluster

First consider some monometallic species, described as organometallic coordination compounds above, in the light of the cluster counting rules. Although the similarity between the isoelectronic 1-CpCo(η^4-B_4H_8) and CpCo(η^4-C_4H_4) is evident in the metal-ligand model, an organometallic chemist would not lightly

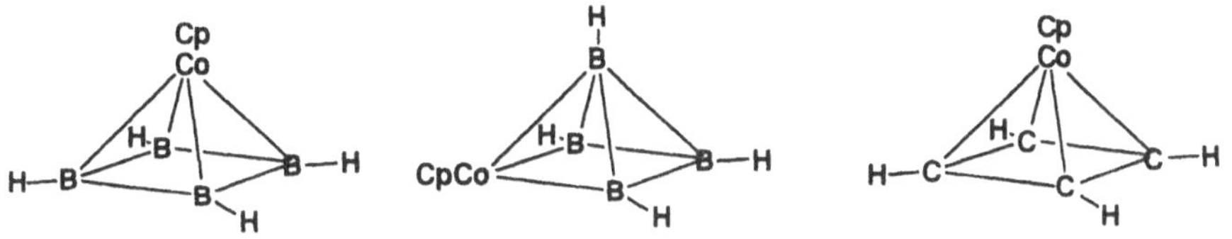

1-$CpCoB_4H_8$ 2-$CpCoB_4H_8$ $CpCoC_4H_4$

Fig. 8. A comparison of the isomeric forms of the *nido*-metallapentaborane $CpCoB_4H_8$ with its isoelectronic organometallic analog $CpCoC_4H_4$. The four BHB bridges of the metallaboranes are not shown

suggest the organometallic analog of the known 2-$CpCoB_4H_8$ (Fig. 8) [78, 79]. The two isomeric forms of the metallaborane are an obvious consequence of the view of these compounds as five-atom heteroclusters rather than as 4-membered planar rings bound to metal centers. Thus, even when the metal-ligand model is perfectly appropriate, the greater flexibility of the cluster view has led to its adoption in the description of new and old metallaboranes, e.g., $(CO)_3MnB_3H_8$ and $(CO)_4MnB_3H_8$ (Fig. 1) can be viewed as 12 electron *nido*- and 14 electron *arachno*-metallatetraboranes analogous with B_4H_8 and B_4H_{10}, respectively [26, 27].

In some cases the metal-ligand model is extremely clumsy at best [80]. Metallaboranes which contain equal, or nearly equal, numbers of main group and transition metal atoms are easily accommodated with the cluster model but not with the metal-ligand model. One example from our own work illustrates the point. A minor product in a complex reaction was isolated and characterized as apparently the first example of a boracyclopropene ring coordinated to a trimetal fragment, i.e., $Fe_3(CO)_9[\eta^3\text{-}B(H)C(H)C(Me)]$ (Fig. 9) [81]. However, if considered as such, it would be a 44-electron unsaturated trimetal complex. There is no geometric evidence nor any other indication that this is an unsaturated compound. Viewed alternatively as a 56-cluster valence electron M_3E_3 saturated cluster, this metallacarborane has the octahedral structure required by the cluster electron counting rules. There are, of course, distortions away

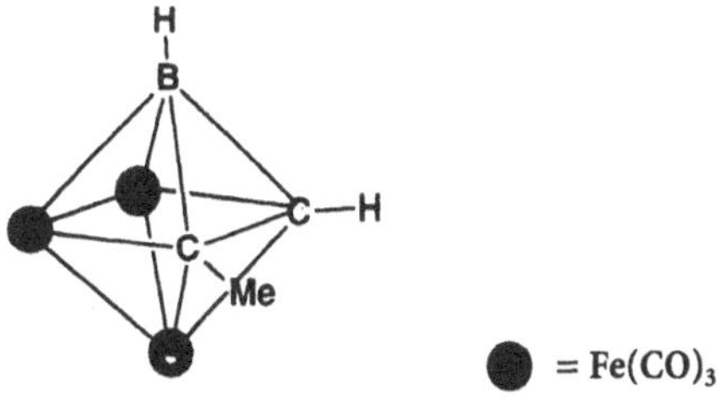

$Fe_3(CO)_9B(H)C(H)C(Me)$

Fig. 9. A metallahexacarborane, $Fe_3(CO)_9B(H)C(H)C(Me)$, with seven cluster bonding pairs and a closo-octahedral cluster structure

from O_h symmetry due to the differing radii of metal and main group atoms. This compound is isolobal with 1,2-$C_2B_4H_6$ and structurally analogous to $Cp_3Co_3B_3H_5$ both of which have seven cluster bonding pairs and are electronically saturated in the sense of following the cluster electron counting rules [82]. Thus, the parent, $Fe_3(CO)_9C_2BH_3$, is only reasonably considered as a six-atom heteroborane.

4.3
Synthetic Ramifications

It has already been suggested above that the electron counting rules, combined with the isolobal principle, stimulated new synthetic chemistry. The efficiency of experimental characterization was boosted. The ability to translate a complex molecular formula into a small set of reasonable structures maximized the usefulness of spectroscopic data. This is true for the characterization of stable species even though crystallographic structure determination often follows. Confidence in spectroscopic characterization is even more important for the intermediate products never isolated. The ready connection between structure and spectra permits reasonable structural assignments. Mechanistic insight, be it only crude and qualitative, takes our understanding of the reaction chemistry to a significantly higher level. It is sometimes forgotten that the complete understanding of the geometric structures of stable species in a system can never give one any real understanding of the time dependence of any given system. Indeed rock-stable molecules are not all that interesting simply because there is not much reaction chemistry associated with them.

One example shows how the interpretation of spectroscopic data permits a better understanding of an overall reaction. The reaction of $[Fe_4(CO)_{12}BH_2]^-$ with $[Rh(CO)_2Cl]_2$ dimer yields only a single final product which was isolated and characterized in the solid state as $[1,6\text{-}Rh_2Fe_4(CO)_{16}B]^-$ [83]. Intermediates in its formation were observed by ^{11}B during the course of the reaction. As is obvious from the spectroscopic observations, one intermediate in its formation is the fragment positional isomer $[1,2\text{-}Rh_2Fe_4(CO)_{16}B]^-$. The identification of this intermediate permitted a study of the mechanism of cluster isomerization from kinetic to thermodynamic product. The mechanism turns out to be one in which soft Lewis bases assist the cluster rearrangement as shown by a rate dependence upon base concentration and properties [84, 85]. The use of the electron counting rules to extrapolate from crystallographically established structures to related intermediates provides superior understanding of the system relative to a single, solid state structure.

The electron counting rules also stimulated synthesis by defining the compositions of potential metallaboranes. The enumeration of the large number of possible compositions (main group and transition metal atoms) and isomers for a given cluster geometry became easy. These sets of molecules served as potential synthetic targets albeit often unconscious ones in the beginning. Some sense of the scope of the chemistry comes from the variations already known. For example, metallaboranes having B_mM_n cluster cores, m and $n \leq 6$, are known for m = 1, n = 1–6; m = 2, n = 1–5; m = 3, n = 1–3; m = 4, n = 1–4; m = 5, n = 1, 4; m = 6, n = 1, 2, i.e., 22 out of the 36 possible combinations. Even though more than one metal fragment type is utilized in these examples it is clear that a wide range of the compounds predicted by the electron counting rules and isolobal principle are accessible.

Given the electron counting rules, it should come as no surprise that mixed-fragment cluster formation is most readily exhibited in practice by fragments that contribute three valence orbitals and two electrons to the cluster bonding system, e.g., BH and the isolobal $Fe(CO)_3$ or CpCo fragments. That is, aggregates of fragments such as CH, and the isolobal $Co(CO)_3$ or CpNi, with three orbitals and three electrons would tend to form chains, rings, and other "three-connect" cages. On the other hand, aggregates of fragments such as BeH, and the isolobal $Mn(CO)_3$ or CpFe, with three orbitals and one electron would tend to form large, extended network structures rather than discrete clusters.

As shown in Fig. 10, sources of metal fragments isolobal to BH have been used to synthesize all isomeric forms of the series of metallaboranes $B_{5-n}H_{9-n}\{ML_x\}_n$, where ML_x is isolobal with BH, for n = 1 and 2. One of three possible forms is known for n = 3 [79, 86–93]. These compounds constitute a pleasing verification of the compositional and isomeric predictions of the electron counting rules. The organometallic view can account for some of the compounds but not all and it does not interrelate the set of compounds.

One must hasten to add that, within limits, a given borane cage can accommodate a large variety of mononuclear main group fragments (BeH, CH, NH,

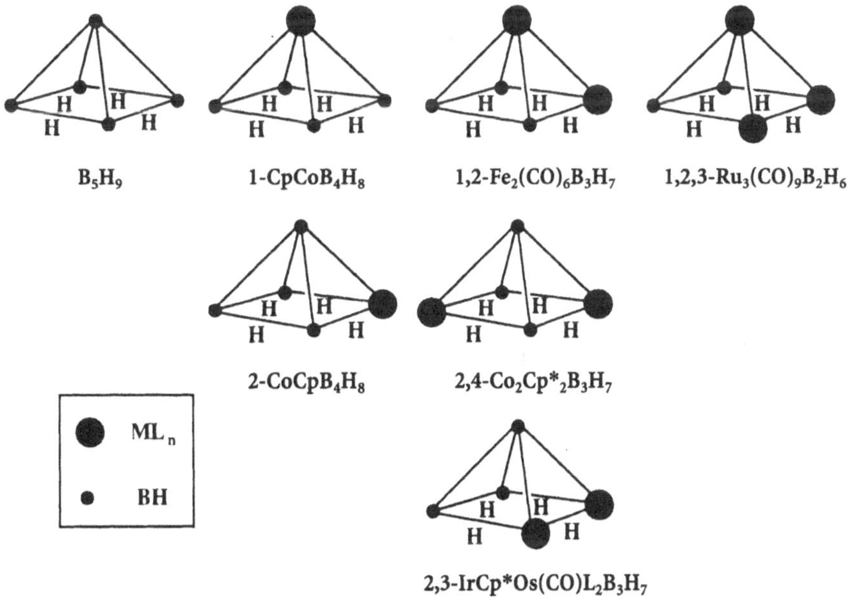

Fig. 10. A set of metallapentaboranes, $B_{5-n}H_{9-n}\{ML_x\}_n$, with seven cluster bonding pairs and a *nido*-square pyramidal geometry

OH) not meeting the three orbital-two electron prescription, as well as the analogous metal fragments, by incorporating net cluster charges or extra or different ligands, e.g., bridging Hs, B-CO instead of B-H. To some extent adjustments are possible with two metal fragments by similar means or, in addition, reducing the number of metal ligands, e.g., a $Co_2(CO)_5$ fragment can be equivalent to two $Fe(CO)_3$ fragments. By the same token, a metal cluster can accommodate a large variety of mononuclear main group atom fragments. An interesting example, characteristic of metal clusters, are single atoms incorporated into the interior of a cluster network. These are considered as interstitial and contribute to the electron count but not to the definition of cluster geometry. So large a variety of main group atoms can be accommodated within a cluster environment that they are discussed in a separate chapter [94].

As already mentioned, it is the intermediate cluster compositions, with two or more metal atoms in a borane cage or two or more main group atoms in a metal cluster, that provide a better test of these ideas. With three BH and three metal fragments there are not so many ways to fiddle with the bonding picture. And, after all, a given metal fragment does not have to be isolobal to a single main group fragment (see the next section). However, in the work to date with fragments such as $Fe(CO)_3$ or CpCo and their heavier congeners, the mixed metal-boron clusters appear generally well behaved relative to the electron counting rules and isolobal principles. Thus, once allowance is made for the differences in radii between boron and the metal, the geometry and placement of the bridging hydrogens in the metallapentaboranes shown in Fig. 10 corresponds directly to that of B_5H_9 itself.

Another interesting example is shown in Fig. 11 where the octahedral cluster geometry appropriate for seven cluster bonding pairs is exhibited as the number of boron atoms ranges from two to four [79, 82, 88, 95, 96]. Curiously, the MB_5 and M_5B clusters are missing from this series of metallaboranes. However, this does not imply instability and their absence is probably due to the present lack of a preparative route.

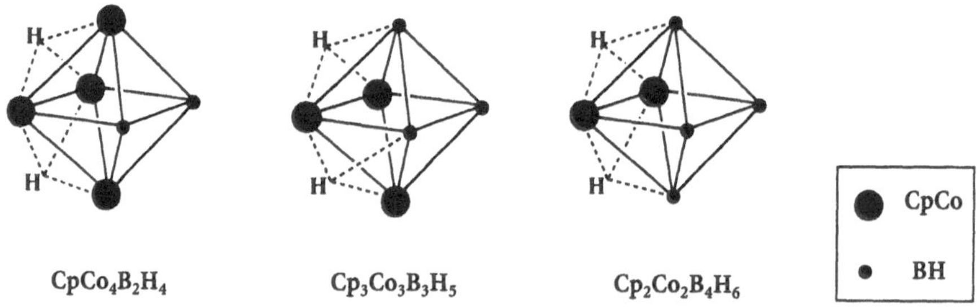

Fig. 11. A set of metallahexaboranes, $Cp_nCo_nB_{6-n}H_{8-n}$, with seven cluster bonding pairs and *closo*-octahedral cluster structures. Note the bridging hydrogens for n=3 are fluxional but are known to be associated with the three Co atoms

4.4 Complications Introduced by the Metal Fragment

There are some very interesting exceptions to and variations on the electron counting rules which are treated more fully in a following chapter [7, 97–105]. These exceptions are often associated with the properties of the metal fragments or a particular dimension of the cluster deltahedron. The $Cp_4M_4B_4H_4$ pair, M = Ni, Co provides an example of an early puzzle associated with the eight-vertex system [102–104]. The four-orbital vs three-orbital metal fragment (isocloso vs hypercloso cluster) debate provides one connected to metal fragment properties [106–109].

It is well documented that (a) the number and type of external ligands on the metal fragment, (b) the dependence of the availability of alternative geometries of similar energy on cluster size, (c) the steric constraints associated with a given cluster geometry [110], (d) the π donor/acceptor capabilities of the ligands [111], (e) the actual electronic role of the so-called "t_{2g}" orbitals on the transition metal center, which are often assumed to be nonbonding in isolobal relationships [16, 100, 112, 113], and (f) the various expressions of the capping principle [77] increase the complexity of the relationship between structure and cluster electron count for transition metal clusters. Of course these complexities encompass more of cluster chemistry than represented by the metallaboranes. The advantage provided by the metallaboranes, assuming that the synthetic chemistry can be developed, is that one can introduce the metal fragments into a cluster in a systematic and controlled fashion. In this manner variations in the isolobal behavior of a metal fragment relative to a borane fragment are more clearly defined.

5 Metallaboranes – The Future

5.1 The Bad News of Successful Models

Successful models lead ultimately to an inability to appreciate the continuous quality of nature. Indeed it seems that the more successful a bonding model the more difficulty chemists have in accommodating systems that lie outside of the capabilities of a model. For example, the 8/18-electron rule plus the idea of a two center-two electron bond is so useful in organizing and discussing so much "general" chemistry that the observed composition of the polyhedral boranes caused considerable conceptual pain in the early days. The discovery of the mere existence of rare gas-halogen compounds generated a burst of surprise in the chemical world which was personally experienced by this author. Both require three center-n electron bond ideas for a minimal description.

The success of the cluster electron counting rules is both described and celebrated in this volume. What, then, do the electron counting rules fail to describe? Exceptions have been touched upon above and systematic variations are treated in detail in a separate chapter. Do any of the exceptions suggest new areas that can be chemically exploited?

The remainder of this chapter explores the use the electron counting rules and the isolobal principle as a foundation from which one seeks new classes of compounds in loose analogy to other established areas. That is, if organic chemistry were limited to hydrocarbons life would be pretty uninteresting. Hydrocarbon-metal complexes (organometallic chemistry) liven things up considerably as do multiple bonds, aromaticity, etc. Although boranes are much more reactive than hydrocarbons and exhibit a significant chemistry, borane-metal complexes (metallaboranes) add additional spice in the same sense that organometallic chemistry does to hydrocarbon chemistry. In the following the idea of electronic unsaturation and multiple bonding in metallaboranes is explored with the aid of recent results in order to illustrate one chemist's view of where the electron counting rules may take us in the future.

5.2
Electronic Unsaturation Applied to Clusters

A simple set of isoelectronic comparisons defines the problem. If one removes two protons and two electrons from C_2H_6 one generates the ethylene or diborane molecule depending on whether the protons are removed from the ligands (hydrogen) or the carbon nuclei. The first is described as unsaturated whereas the latter requires a three center-two electron bond to describe the bridging interaction. As shown in Fig. 12a, the same exercise with benzene yields benzyne (multiple bond) or the nido-$C_4B_2H_6$ carborane cage (multicenter bonding). With a metal dimer an alternative outcome is possible (Fig. 12b). For example, removal of 2 CO (equivalent to 4H) from $[CpFe(CO)_2]_2$ yields a dimer, presumably with a triple bond [114], whereas removal of two protons from each of the metal nuclei and four electrons yields $[CpCr(CO)_2]_2$ also with a triple bond. As pointed out by Hoffmann [16], the $CpCr(CO)_2$ frag-

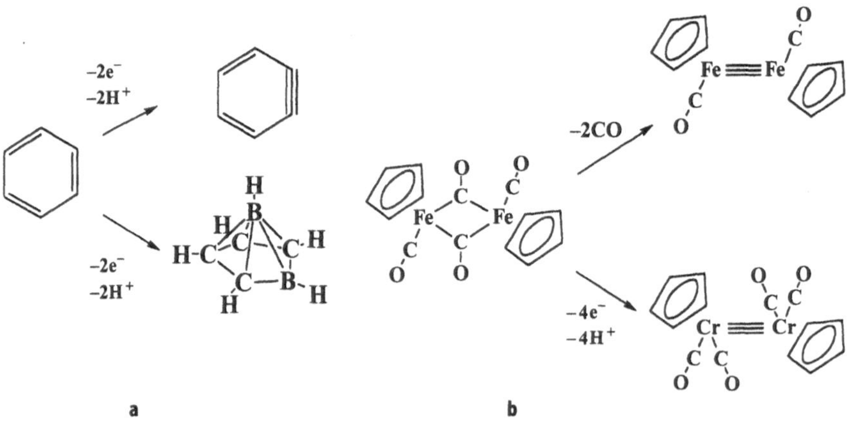

Fig. 12a, b. Contrasting responses of main group and transition element systems

ment can combine two of its "t_{2g}" orbitals with the normal valence orbital of a $CpM(CO)_2$ fragment to generate a three orbital-3-electron fragment isolobal with CH. Not only do these orbitals possess the proper symmetry but they also lie at an appropriate energy for bonding. Of course, an organometallic chemist would say that a dimer formed from 15 electron $[CpCr(CO)_2]$ fragments requires a triple bond to satisfy the 18-electron rule but this rationale avoids the essence of the problem.

The answer is not known if one does the same mental exercise with a dimetallaborane cluster, e.g., go from the known nido-$Fe_2(CO)_6B_3H_7$ [89] analog of B_5H_9 to the hypothetical $Cr_2(CO)_6B_3H_7$ with no change in the ancillary ligands on the metal fragments. Will the cluster become more condensed due to a decrease in cluster bonding electrons? Will a localized M-M multiple bond form? Will an electronically delocalized unsaturated cluster be formed or will there be some response that is unanticipated?

What, then, is unsaturation in the cluster context? The meaning of unsaturation in general is a rather subtle one and its application to clusters is not straightforward. Relative to an electron precise cage, e.g., P_4, a cluster such as $[B_6H_6]^{2-}$, might be viewed as unsaturated in the sense that the number of cluster bonding pairs (7) is less than the number of deltahedral edges (12). For clusters, electronic unsaturation will be associated with fewer cluster bonding electrons than required for the observed core geometry by the electron counting rules. Thus, geometry is still the first source of information on bonding but can be ambiguous in a cluster context particularly if the isolobal relationships themselves are no longer unique. Hence, theory above the level of counting electrons is a necessary component in discussions of electronic structure. In addition, the reactivity of the new molecules can provide essential corroboration of models based on geometry and calculations.

5.3 Unsaturated Metallaboranes

Recently we have reported an example of a new type of metallaborane, i.e., the chromaborane, $Cp^*_2Cr_2B_4H_8$ [115, 116]. A few mononuclear chromacarboranes and boranes have been reported [9, 35, 117–119], but this dinuclear species is essentially different and serves as an example of the kind of compounds that may be important in the future.

The geometry of the diamagnetic $Cp^*_2Cr_2B_4H_8$ cluster (Fig. 13a) can be described as an apparent nido framework based on a pentagonal bipyramidal deltahedron with an unoccupied equatorial vertex. The Cr-Cr distance of 2.87 Å clearly rules out multiple bonding but is ambiguous concerning the existence of a bonding interaction. It is longer than the 2.72 Å distance in $[Cp^*CrH]_4$ [120, 121], which is characterized as nonbonding, but shorter than the 3.28 Å distance in $[Cp^*Cr(CO)_3]_2$ which is bonding, albeit weakly. However, a fragment MO analysis (Fig. 13b) points to a well defined filled Cr-Cr bonding orbital, generating a positive overlap population, which is paired with an equally well defined, unfilled Cr-Cr antibonding orbital. The origin of these orbitals lies in a component of the "t_{2g}" set of the Cp*Cr fragment. Indeed, the calculations show that the "t_{2g}" set

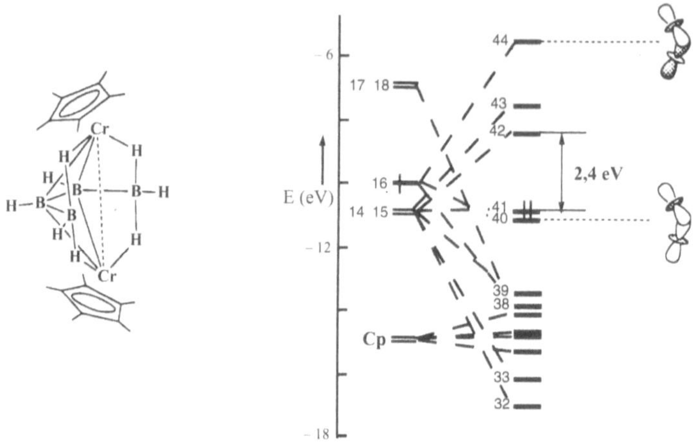

Fig. 13. The geometric structure of $Cp^*_2Cr_2B_4H_8$ and its MO structure showing the Cr-Cr bonding and antibonding orbitals and their origin from the CpCr fragments

participates not only in Cr-Cr bonding but also in Cr-B cluster bonding. The net result is a strong interaction between the orbitals of both metal and borane fragments to give a closed shell molecule with a HOMO energy and HOMO-LUMO gap typical of metallaboranes containing later transition metals.

$Cp^*_2Cr_2B_4H_8$ does not obey the electron counting rules. Consider the structure of $Cp_2Co_2B_2H_2S_2$ in Fig. 14a (d_{CoCo}=3.07 Å) [122]. This analog of the presently unknown $Cp_2Co_2B_4H_8$ cluster obeys the Wade-Mingos definition for a *nido* six-atom cluster with 48 cluster valence electrons (cve) and a nonbonding CoCo interaction. $Cp^*_2Cr_2B_4H_8$ displays a similar geometry but only possesses 42 cve, apparently 6 short of the required number. In light of the Cr-Cr bonding revealed by the MO calculations, a better geometric description of $Cp^*_2Cr_2B_4H_8$ is that of a Cr_2B_2 tetrahedron capped by two BH_3 units which requires 44 cve. Although this is a better match, $Cp^*_2Cr_2B_4H_8$ is still short two electrons from

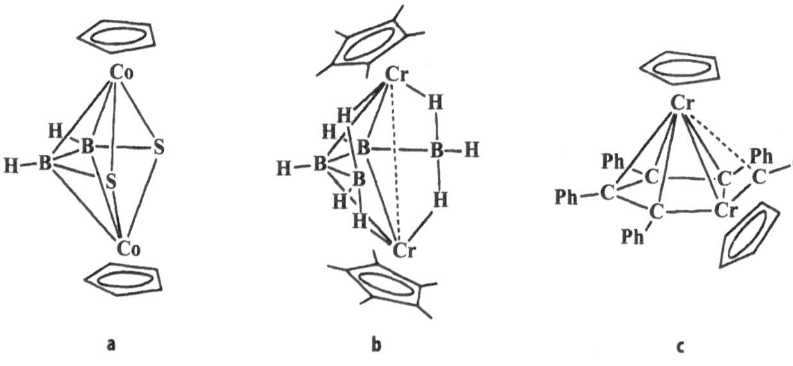

Fig. 14a–c. Comparison of: **a** $Cp_2Co_2B_2H_2S_2$; **b** $Cp^*_2Cr_2B_4H_8$; **c** $Cp_2Cr_2(CO)C_4Ph_4$

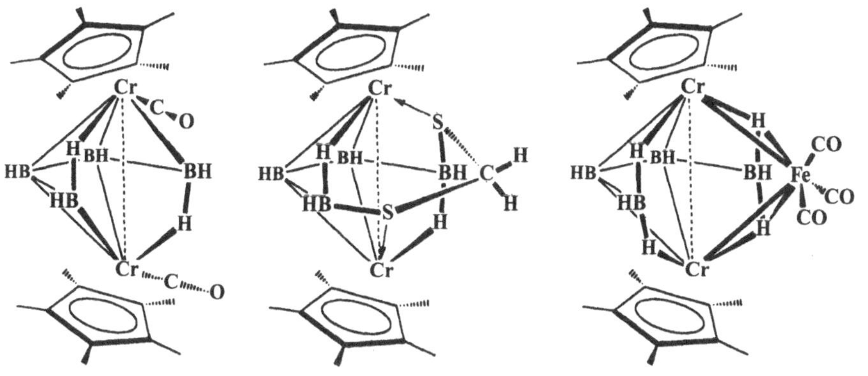

a b c

Fig. 15a–c. Structures of the products of the reaction of $Cp^*{}_2Cr_2B_4H_8$ with: **a** CO; **b** CS_2; **c** $Fe_2(CO)_9$

the prescribed number. A closely related organometallic cluster, $Cp_2Cr_2(CO)C_4Ph_4$ [123], adopts a true *nido* geometry based on a pentagonal bipyramidal deltahedron with an axial vertex vacant (Fig. 14c). This geometry requires 48 cve but only 44 cve are available. It, too, is electronically unsaturated but this unsaturation is expressed in a very short Cr-Cr interaction that the authors describe as a triple bond. We conclude that $Cp^*{}_2Cr_2B_4H_8$ (Fig. 14b) is electronically unsaturated in the same sense but that this unsaturation is delocalized throughout the cluster bonding network.

This description of $Cp^*{}_2Cr_2B_4H_8$ is corroborated by its reactivity. A facile reaction with CO produces $Cp^*{}_2Cr_2(CO)_2B_4H_6$ (Fig. 15a) with a core structure qualitatively similar to that of $Cp^*{}_2Cr_2B_4H_8$ but with the prescribed 44 cve of a saturated cluster [116]. Reaction with CS_2 also leads to addition but now both CS double bonds have been reduced to form a cluster bound H_2CS_2 ligand, i.e., the $Cp^*{}_2Cr_2B_4H_8S_2CH_2$ molecule (Fig. 15b) with 48 cve [124]. Again, the Cr_2B_4 core structure is qualitatively unchanged from that of $Cp^*{}_2Cr_2B_4H_8$.

Each compound exhibits a satisfactory MO diagram (HOMO-LUMO gap, charges, etc.) and it appears clear that the Cp*Cr fragment adjusts its interactions relative to the demands of the cluster bonding network. Although there are examples of a single cluster structure with two different electron counts, to the best of our knowledge this constitutes the first example of a cluster geometry with three (four if the Co analog is included) different electron counts. (There are small, but significant, geometric changes with electron count, which correlate with the frontier MO bonding/antibonding characters, but space does not permit discussion.) Whatever the detailed electronic structure, the empirical facts are that $Cp^*{}_2Cr_2B_4H_8$ is an unsaturated cluster (Lewis acidic) with hydride reducing equivalents available in adjacent cluster sites (Lewis basic). It is important to note in this regard that the chromaborane is significantly different from a typical ferraborane. The latter are saturated and require more activation for reaction. The ferraboranes also possess hydrogens that are much more protonic in character and many ferraboranes are stable to aqueous acid.

The interesting nature of $Cp^*{}_2Cr_2B_4H_8$ is further revealed in a reaction with $Fe_2(CO)_9$ to yield $Cp^*{}_2Cr_2B_4H_8Fe(CO)_3$ (Fig. 15c) [125]. The simplest description of the structure (supported by an MO fragment analysis) is that of a 4-electron $Cp^*{}_2Cr_2B_4H_8$ ligand bound to an $Fe(CO)_3$ fragment. That is, two Fe-Cr single bonds are formed utilizing two electrons each from $Cp^*{}_2Cr_2B_4H_8$ and $Fe(CO)_3$ and then a BH_2 unit chelates the Fe atom in the manner of a borohydride ligand. $Cp^*{}_2Cr_2B_4H_8$ picks up two electrons and satisfies the electron counting rules (44 cve) and the $Fe(CO)_3$ picks up four electrons so that the Fe atom satisfies the 18 electron rule. In contrast to the reaction with CS_2 the hydrides are not transferred but only bridge between boron and iron atoms.

There is a fundamental question concerning the origin of the delocalized unsaturation in $Cp^*{}_2Cr_2B_4H_8$. It may well arise in the steric demands of the Cp* ligands in that they prevent the close approach required for the formation of a metal-metal multiple bond such as that found in the organometallic compound in Fig. 14c. In order to investigate this possibility the Cp analog, presently unknown, must be prepared. Of course, the origin of the delocalized unsaturation may be essentially electronic and associated with the real difference between boron and carbon. These and other interesting questions have yet to be answered; however, it seems clear that metallaborane electronic unsaturation, introduced by the presence of two or more early transition metal fragments, provides an unanticipated area for future research. Though unanticipated by the electron counting rules, it is only clearly seen with the perspective provided by them.

Acknowledgements. I thank my coworkers, listed in the references, for their excellent efforts in carrying out the chemistry that originated at Notre Dame and I thank the National Science Foundation for the financial support of this work.

6 References

1. Fehlner TP (1992) Inorganometallic chemistry. Plenum, New York
2. Elschenbroich C, Salzer A (1989) Organometallics. VCH, New York
3. Housecroft CE, Fehlner TP (1982) Adv Organomet Chem 21:57
4. Housecroft CE (1992) In: Fehlner TP (ed) Inorganometallic chemistry. Plenum, New York, p73
5. Kennedy JD (1984) Prog Inorg Chem 32:519
6. Kennedy JD, (1986) Prog. in Inorg. Chem. 34:211
7. Grimes RN, (1982) Pure & Appl. Chem. 54:43
8. Grimes RN (1982) In: Grimes RN (ed) Metal interactions with boron clusters. Plenum, New York, p269
9. Hawthorne MF (1975) J Organomet Chem 100:97
10. Hawthorne MF (1972) Pure Appl Chem 29:547
11. Fehlner TP (1990) Adv Inorg Chem 35:199
12. Williams RE (1971) Inorg Chem 10:210
13. Wade K (1972) Inorg Nucl Chem Lett 8:559
14. Rudolph RW (1976) Acc Chem Res 9:446
15. Mingos DMP (1972) Nature (London) Phys Sci 236:99
16. Hoffmann R (1981) Science 211:995
17. Campagna S, Giannetto A, Serroni S, Denti G, Trusso S, Mallamace F, Micali N (1995) J Am Chem Soc 117:1754

18. Park JW, Mackenzie PB, Schaefer WP, Grubbs RH (1986) J Am Chem Soc 108:6402
19. Jones WD, Feher FJ (1986) J Am Chem Soc 108:4814
20. Graham WAG (1986) J Organomet Chem 300:81–91
21. Bergman RG, Seidler PF, Wenzel TT (1985) J Am Chem Soc 107:4358
22. Crabtree RH, Holt EM, Lavin M, Morehouse SM (1985) Inorg Chem 24:1986
23. Hoyano JK, McMaster AD, Graham WAG (1983) J Am Chem Soc 105:7190
24. James BD, Wallbridge MGH (1970) Prog Inorg Chem 11:99
25. Marks TJ, Kolb JR (1977) Chem Rev 77:263
26. Gaines DF, Hildebrandt SJ (1978) Inorg Chem 17:794
27. Hildebrandt SJ, Gaines DF, Calabrese JC (1978) Inorg Chem 17:790
28. Calcbrese JC, Fischer MB, Gaines DF, Lott JW (1974) J Am Chem Soc 96:6318
29. Baker RT et al. (1984) J Am Chem Soc 106:2965–3025
30. Behnken PE, Hawthorne MF (1984) Inorg Chem 23:3420
31. Behnken PE et al. (1984) J Am Chem Soc 106:7444
32. Belmont JA, Soto J, King RE III, Donaldson AJ, Hewes JD, Hawthorne MF (1989) J Am Chem Soc 111:7475
33. Jeffery JC, Jelliss PA, Stone FGA (1994) Organometallics 13:2651
34. Snow SA, Shimoi M, Ostler CD, Thompson BK, Kodama G, Parry RW (1984) Inorg Chem 23:511
35. Shimoi M, Katoh K, Ogino H (1990) J Chem Soc Chem Commun 811
36. Katoh K, Shimoi M, Ogino H (1992) Inorg Chem 31:670
37. Aldridge S, Blake AJ, Downs AJ, Parsons S (1995) J CS Chem Commun 1363
38. Ivanov SV, Lupinetti AJ, Miller SM, Anderson OP, Solntsev KA, Strauss SH (1995) Inorg Chem 34:6419
39. Greenwood NN, Ward IM (1974) Chem Soc Rev 3:231
40. Hawthorne MF, Young DC, Wegner PA (1965) J Am Chem Soc 87:1818
41. Siebert W (1981) In: Müller A, Diemann E (eds) Transition metal chemistry. Verlag Chemie, Deerfield Beach, Florida p158
42. Siebert W (1985) Z Naturforsch B:Anorg Chem Org Chem 40B:458
43. Grimes RN (1992) Chem Rev 92:251
44. Hawthorne MF, Dunks GB (1972) Science 178:462
45. Warren LF Jr, Hawthorne MF (1970) J Am Chem Soc 92:1157
46. Wang X, Sabat M, Grimes RN (1995) J Am Chem Soc 117:12227
47. Crowther DJ, Borkowsky SL, Swenson D, Meyer TY, Jordan RF (1993) Organometallics 12:2897
48. Stockman KE, Sabat M, Finn MG, Grimes RN (1992) J Am Chem Soc 114:8733
49. Grimes RN (1979) Coordination Chemistry Reviews 28:47
50. Coffy TJ, Medford G, Plotkin J, Long GJ, Huffman JC, Shore SG (1989) Organometallics 8:2404
51. DeKock RL, Deshmukh P, Fehlner TP, Housecroft CE, Plotkin JS, Shore SG (1983) J Am Chem Soc 105:815
52. Aradi AA, Fehlner TP (1990) Adv Organomet Chem 30:189
53. Jan D-Y, Hsu L-Y, Workman DP, Shore SG (1987) Organometallics 6:1984
54. Jan D-Y, Shore SG (1987) Organometallics 6:428
55. Jan D-Y, Workman DP, Hsu L-Y, Krause JA, Shore SG (1992) Inorg Chem 31:5123
56. Fehlner TP (1995) In: Farrugia LJ (ed) The synergy between dynamics and reactivity at clusters and surfaces. Kluwer Academic, Dordrecht, p75
57. Dutta TK, Vites JV, Jacobsen GB, Fehlner TP (1987) Organometallics 6:842
58. Lynam MM, Chipman DM, Barreto RD, Fehlner TP (1987) Organometallics 6:2405
59. Vites JC, Eigenbrot C, Fehlner TP (1984) J Am Chem Soc 106:4633
60. Vites JC, Housecroft CE, Jacobsen GB, Fehlner TP (1984) Organometallics 3:1591
61. Vites JC, Housecroft CE, Eigenbrot C, Buhl ML, Long GJ, Fehlner TP (1986) J Am Chem Soc 108:3304
62. Fehlner TP (1990) Polyhedron 9:1955
63. Barreto RD, Puga J, Fehlner TP (1990) Organometallics 9:662

64. Shore SG (1975) In: Muetterties EL (ed.) Boron hydride chemistry. Academic Press, New York, p79
65. Williams RE (1976) Adv Inorg Chem & Radiochem 18:67
66. Moore EJ, Sullivan JM, Norton JR (1986) J Am Chem Soc 108:2257
67. Walker HW, Pearson RG, Ford PC (1983) J Am Chem Soc 105:1179
68. Kristjánsdóttir SS, Moody AE, Weberg RT, Norton JR (1988) Organometallics 7:1983
69. Weberg RT, Norton JR (1990) J Am Chem Soc 112:1105
70. Jun C-S, Fehlner TP (1994) Organometallics 13:2145
71. Jun C-S, Halet J-F, Rheingold AL, Fehlner TP (1995) Inorg Chem 34:2101
72. Jun C-S, Bandyopadhyay AK, Fehlner TP (1996) Inorg Chem 35:2189
73. Wade K (1974) New Scientist 62:615
74. Wade K (1976) Adv Inorg Chem & Radiochem 18:1
75. Mingos DMP (1984) Acc Chem Res 17:311–319
76. Mingos DMP, Johnston RL (1987) Structure and Bonding 68:29
77. Mingos DMP, Wales DJ (1990) Introduction to cluster chemistry. Prentice Hall, New York
78. Venable TL, Sinn E, Grimes RN (1984) J Chem Soc, Dalton Trans 2275
79. Weiss R, Bowser JR, Grimes RN (1978) Inorg Chem 17:1522
80. Housecroft CE, Fehlner TP (1982) Inorg Chem 21:1739
81. Meng X, Fehlner TP, Rheingold AL (1990) Organometallics 9:534
82. Pipal JR, Grimes RN (1977) Inorg Chem 16:3255
83. Khattar R, Puga J, Fehlner TP, Rheingold AL (1989) J Am Chem Soc 111:1877
84. Bandyopadhyay AK, Khattar R, Fehlner TP (1989) Inorg Chem 28:4434
85. Bandyopadhyay AK, Khattar R, Puga J, Fehlner TP, Rheingold AL (1992) Inorg Chem 31: 465
86. Lipscomb WN (1963) Boron hydrides. Benjamin, New York
87. Greenwood NN, Savory CG, Grimes RN, Sneddon LG, Davison A, Wreford SS (1974) JCS Chem Comm 718
88. Miller VR, Weiss R, Grimes RN (1977) J Am Chem Soc 99:5646
89. Andersen EL, Haller KJ, Fehlner TP (1979) J Am Chem Soc 101:4390
90. Nishihara Y, Deck KJ, Shang M, Fehlner TP, Haggerty BS, Rheingold AL (1994) Organometallics 13:4510
91. Bould J, Pasieka M, Braddock-Wilking J, Rath NP, Barton L, Gloeckner C (1995) Organometallics 14:5138
92. Chipperfield AK, Housecroft CE, Matthews DM (1990) J Organomet Chem 384:C38
93. Housecroft CE, Matthews DM, Rheingold AL (1992) JCS Chem Comm 323
94. Shriver DF, Kaesz HD, Adams RD (1990) The chemistry of metal cluster complexes. VCH, New York
95. Feilong J, Fehlner TP, Rheingold AL (1987) J Am Chem Soc 109:1860
96. Pipal JR, Grimes RN (1979) Inorg Chem 18:252
97. O'Neill ME, Wade K (1982) Inorg Chem 21:461
98. Cox DN, Mingos DMP, Hoffmann R (1981) J Chem Soc, Dalton Trans 1788
99. Fowkes H, Greenwood NN, Kennedy JD, Thornton-Pett M (1986) J Chem Soc, Dalton Trans 517
100. Halet J-F, Hoffmann R, Saillard J-Y (1985) Inorg Chem 24:1695
101. Vahrenkamp H, Walter D (1982) Organometallics 1:874
102. Pipal JR, Grimes RN (1979) Inorg Chem 18:257
103. Bowser JR, Grimes RN (1978) J Am Chem Soc 100:4623
104. Bowser JR, Bonny A, Pipal JR, Grimes RN (1979) J Am Chem Soc 101:6229
105. Jun CS, Fehlner TP, Rheingold AL (1993) J Am Chem Soc 115:4393
106. Bould J, Greenwood NN, Kennedy JD, McDonald WS (1982) J Chem Soc Chem Commun 465
107. Kennedy JD (1986) Inorg Chem 25:11
108. Baker RT (1986) Inorg Chem 25:109
109. Johnson RL, Mingos DMP (1986) Inorg Chem 25:3321
110. Mingos DMP (1982) Inorg Chem 21:464

111. Chisholm MC, Clark DL, Hampden-Smith MJ, Hoffman DH (1989) Angew Chem Int Ed Engl 28:432
112. Mingos DMP (1985) J Chem Soc, Chem Commun 1352
113. Halet J-F, Saillard J-Y (1987) New J Chem 11:315
114. Kvietok FA, Bursten BE (1994) J Am Chem Soc 116:9807
115. Deck KJ, Nishihara Y, Shang M, Fehlner TP (1994) J Am Chem Soc 116:8408
116. Ho J, Deck KJ, Nishihara Y, Shang M, Fehlner TP (1995) J Am Chem Soc 117:10292
117. Maynard RB, Wang Z-T, Sinn E, Grimes RN (1983) Inorg Chem 22:873
118. Spencer JT, Grimes RN (1987) Organometallics 6:323,328,335
119. Oki A et al. (1992) Organometallics 11:4202
120. Heintz RA, Haggerty BS, Wan H, Rheingold AL, Theopold KH (1992) Angew Chem Int Ed Engl 31:1077
121. Heintz RA, Ostrander RL, Rheingold AL, Theopold KH (1994) J Am Chem Soc 116:11387
122. Micciche RP, Carroll PJ, Sneddon LG (1985) Organometallics 4:1619
123. Knox SAR, Stansfield RFD, Stone FGA, Winter MJ, Woodward P (1982) J Chem Soc, Dalton Trans 173
124. Hashimoto H, Shang M, Fehlner TP (1996) Organometallics 15:1963
125. Hashimoto H, Shang M, Fehlner TP (1996) J Am Chem Soc 118:8164

Clusters with Interstitial Atoms from the *p*-Block: How do Wade's Rules Handle Them?

Catherine E. Housecroft

Institut für Anorganische Chemie, Universität Basel, Spitalstrasse 51, CH-4056 Switzerland.
E-mail: housecroft@ubaclu.unibas.ch

This article is dedicated to Ken Wade whose thought-provoking undergraduate lectures provided inspiration at a time when Wade's rules were breaking new ground.

This article considers the application of Wade's rules of cluster electron counting to low oxidation state transition metal clusters containing interstitial atoms from the *p*-block. The discussion focuses on the classification of clusters using Wade's rules, and on their successes and failures which can be rationalized in terms of the relationship of a particular metal cluster to a deltahedral $[B_nH_n]^{2-}$ anionic prototype cluster. The need to extend the electron counting schemes to the Polyhedral Skeletal Electron Pair Theory (PSEPT) is seen when we consider frameworks such as the trigonal prism which is considered as a 3-connected polyhedron rather than as a *hypho*-cluster. A square-antiprism M_x-skeleton is ambiguous: such clusters may or may not obey Wade's rules. The application of Wade's rules to condensed polyhedra and the limitations of the rules in this case are considered. Finally, we discuss examples of gold clusters containing interstitial atoms from the *p*-block and illustrate that the bonding in and the shapes of these clusters are not satisfactorily rationalized by either the Wade or PSEPT approach.

1 Introduction

The use of Wade's rules [1] for the rationalization or prediction of borane and carbaborane clusters as well as other heteroboranes has been of prime importance to main group chemists. In 1975, in a feature article in *Chemistry in Britain*, Wade considered their application to clusters in a more general sense and showed their relevance to several metal carbonyl clusters including the carbides *closo*-$Ru_6(CO)_{17}C$ and *nido*-$Fe_5(CO)_{15}C$ [2]. At this point in the development of cluster chemistry, the number of species containing interstitial atoms was small, providing only a very limited 'test sample' for Wade's electron counting schemes. Over the past two decades, the number of low-oxidation state transition metal cluster compounds reported in the chemical literature has increased enormously and among this class of compound is a growing sub-class of clusters containing interstitial atoms from the *p*-block.

The shapes of *low* nuclearity metal cluster cores are often readily rationalized in terms of Wade's rules in which an *n* vertex closed *deltahedral* frame is taken to require $(n + 1)$ pairs of bonding electrons. However, for higher nuclearity and condensed clusters, consideration of the total valence electron count has proved valuable [3, 4]. In this article, we consider examples in transition metal carbonyl chemistry taken from the last ten years or so, focusing on compounds which contain atoms from the *p*-block in semi- or fully-interstitial environments (Fig. 1). We consider to what extent these clusters obey or challenge Wade's electron counting rules. It is not the intention to reiterate basic theories; such discussions, including the isolobal relationships between borane and metal carbonyl fragments (summarized in part in Fig. 2), can be found elsewhere [1–7].

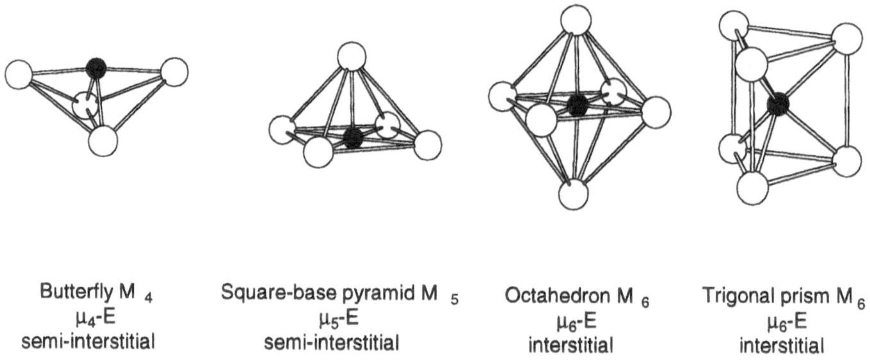

Fig. 1. Four modes of bonding for semi- and fully-interstitial *p*-block atoms in low oxidation state metal carbonyl (and related) clusters which are well represented in the literature

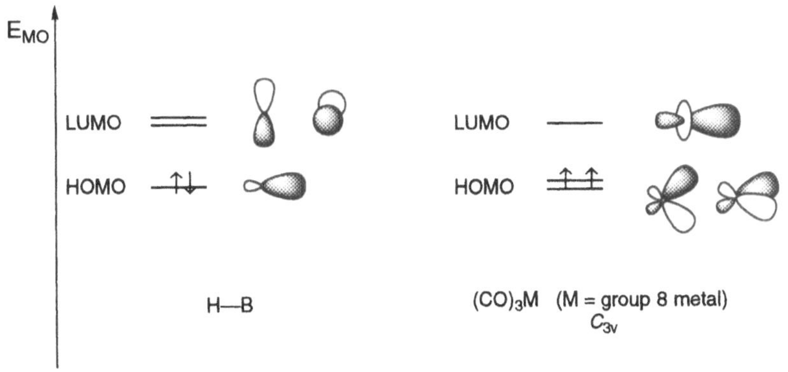

Fig. 2. An example of the isolobal principle at work: the BH unit which is a fundamental building block of borane and heteroborane clusters is isolobal with a C_{3v} $M(CO)_3$ unit (M is Fe, Ru or Os) which is an important building block in metal carbonyl cluster chemistry

2
The Octahedral Cavity and Capped Octahedra

We begin this discussion by looking at the structures of representative carbido, borido and nitrido-species, a good number of which are 'well-behaved' so far as Wade's electron counting schemes are concerned. The octahedral cavity has been a common motif in carbide chemistry and, more recently, borides and nitrides have also been established. (We shall look later at interstitial cavities suitable for large *p*-block atoms). A cluster such as $Ru_6(CO)_{17}C$ is the counterpart of $[B_6H_6]^{2-}$ – both possess seven electron pairs for cluster bonding and both are *closo*-species.

A problem that we must recognize for a carbonyl cluster is the spatial requirements of the ligands. Whilst this is moderate for the CO ligand, it is significant for Lewis bases such as phosphines, R_3P. The interstitial atom can provide electrons from *within* the cage; an interstitial carbon atom effectively 'replaces' two 2-electron donor ligands whilst adding nothing to the steric crowding around the metal framework. The packing of the carbonyl ligands in $Ru_6(CO)_{17}C$ [8] is illustrated in Fig. 3; a related species with an empty Ru_6-cavity would have the formula '$Ru_6(CO)_{19}$' and would be sterically congested (see below, however, for a related rhenium species). Alternatively, the cluster can bear one or more hydrogen atoms or carry a negative charge, for example *closo*-$[Ru_6(CO)_{18}]^{2-}$ [9] but the presence of an interstitial atom precludes the necessity of the charge becoming excessively high.

By using Wade's approach, one can predict a wide range of possible octahedral carbide, boride or nitride species; each cluster requires seven pairs of skeletal electrons, and, with the restriction that all the ligands are carbonyls, Table 1 shows fragments that might be combined to give a Wadean cluster containing an interstitial boron, carbon or nitrogen centre. Of course, putting the fragments together on paper is a very different matter from combining them in

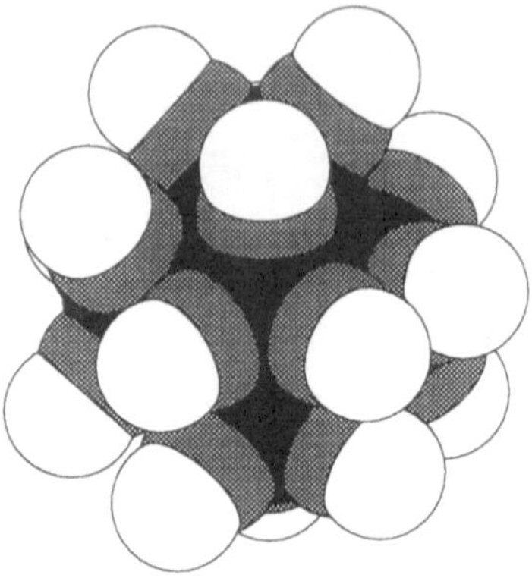

Fig. 3. Space filling diagram of the solid state structure of $Ru_6(CO)_{17}C$

practice, and synthetic routes often yield a range of products, some of which are frequently unexpected. The later the periodic group in which the metal resides, the fewer ligands are needed to attain the required number of cluster bonding electrons – contrast $[Re_6(CO)_{19}C]^{2-}$ [10] with $[Co_6(CO)_{13}C]^{2-}$ [11], both of which are *closo*-octahedral carbides.

Replacement of carbonyl ligands by organic fragments provides members of a class of cluster often cited as models for surface species. An η^6-C_6H_6 ligand provides six electrons; in Table 1, $M(\eta^6$-$C_6H_6)$ could replace an $M(CO)_3$ unit, and other aromatics may behave similarly. This is nicely demonstrated in the carbide $Ru_6(CO)_{14}(\eta^6$-$C_6H_5Me)C$ [12] (Fig. 4) which is an analogue of $Ru_6(CO)_{17}C$ [8]. The crystal structure of $Ru_6(CO)_{12}(\eta^6$-$C_6H_5Me)(\mu$:η^2:η^2-$C_6H_8)C$ [13] (Fig. 4) confirms the presence of both a 6-electron benzene unit and a 4-electron cycloocta-1,3-diene, illustrating the way in which Wade's theories can be extended to alkenes as well as aromatics.

Wade's rules deal with capped clusters by allowing the number of skeletal atoms to be extended (as caps over triangular faces) without the electron count being increased. In a metal cluster, the justification for this rule lies in the fact that the frontier orbitals of the capping ML_x unit interact with orbitals that are directed out of an M_3-triangular face. Such orbital interactions do not disrupt the bonding in the cluster core and do not alter the number of cluster molecular orbitals that must be occupied. Each of $[Re_6(CO)_{19}C]^{2-}$ [10], $[Re_7(CO)_{22}C]^{-}$ [14]

Table 1. Electrons available for cluster bonding from $M(CO)_x$ fragments; interstitial atoms provide *all* their valence electrons for cluster bonding

Cluster fragment	M = group 6 metal	M = group 7 metal	M = group 8 metal	M = group 9 metal	M = group 10 metal
$M(CO)_2$	−2	−1	0	1	2
$M(CO)_3$	0	1	2	3	4
$M(CO)_4$	2	3	4	5	6

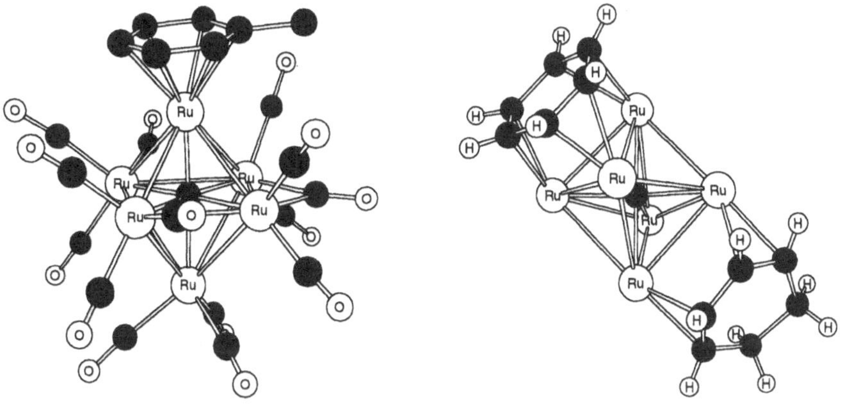

Fig. 4. Solid state structures of $Ru_6(CO)_{14}(\eta^6\text{-}C_6H_5Me)C$ and $Ru_6(CO)_{12}(\eta^6\text{-}C_6H_5Me)(\mu\text{:}\eta^2\text{:}\eta^2\text{-}C_6H_8)C$

and $[Re_7(CO)_{21}C]^{3-}$ [15] possesses seven electron pairs (Table 1) and each has an octahedral core, but in the Re_7-anions, the core is capped on one face by a rhenium fragment (Fig. 5a). This family of clusters provides us with an opportunity to investigate how the presence of gold(I) phosphine fragments is treated by electron counting rules.

Figure 5b shows the core structure of the carbide $[Re_7(CO)_{21}(AuPPh_3)C]^{2-}$ [16]. In this anion, the $AuPPh_3$ unit adopts a capping position, but in some species including a number with interstitial atoms, for example in $[Ru_6(CO)_{17}\{AuP(C_6H_4\text{-}2\text{-}Me)_3\}B]^{2-}$ [17], the gold(I) phosphine unit is in an edge-bridging site. The face-capping and edge-bridging sites are similarly adopted by cluster-bound hydrogen atoms, and the isolobal relationship between H^+ and $[AuPR_3]^+$ (or H and $AuPR_3$) is now widely accepted amongst the cluster community [18, 19]. Despite the differing steric requirements of the two units, their similar frontier orbital properties permit the bonding sites to mimic

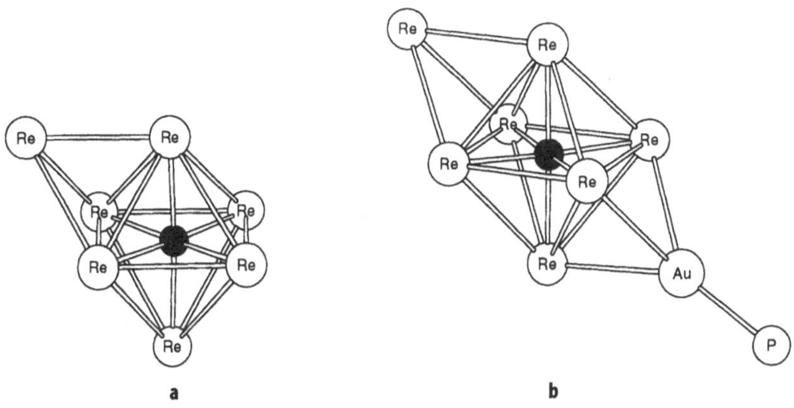

Fig. 5 a, b. Structures of anions: **a** $[Re_7(CO)_{22}C]^-$; **b** $[Re_7(CO)_{21}(AuPPh_3)C]^{2-}$

one another in many instances. There are of course exceptions, the most common being when multiple gold(I) fragments are introduced and Au-Au interactions become important. An $[AuPR_3]^+$ unit then provides no electrons for cluster bonding, and capping (or edge bridging) units can be added to a cluster anion without increasing the number of skeletal electron pairs. The transformation from $[Re_7(CO)_{21}C]^{3-}$ to $[Re_7(CO)_{21}(AuPPh_3)C]^{2-}$ involves no structural change of the central Re_7C-core and both anions are capped *closo*-clusters by Wade's rules. The gold centre of an $[AuPR_3]$ unit is *not* considered to be a vertex atom in the metal polyhedron but instead $[AuPR_3]$ is regarded as being a peripheral unit having no more electronic influence than a proton. We return to gold-containing species in Sect. 8.

Returning to the basic capping principle, some of the most beautiful examples of the rule at work are anions containing an M_{10}-cage in which four faces of an octahedron are capped. Figure 6 shows the structures of $[HOs_{10}(CO)_{24}C]^-$ [20] and $[Ru_{10}(CO)_{24}N]^-$ [21] along with a schematic representation of the metal core in which the four capping atoms are highlighted. In $[HOs_{10}(CO)_{24}C]^-$, the fragments can be partitioned into ten $Os(CO)_2$ units, one H and four CO ligands, a C atom and a negative charge which provide seven electron pairs consistent with an octahedral cage. Wade's rules predict that of the ten osmium centres, six combine to give the parent *closo*-cluster and four adopt capping positions. Notice that the formal partitioning of the cluster units for the purposes of electron counting has no bearing, necessarily, on the experimentally observed arrangement of the ligands. The same electron count is obtained no

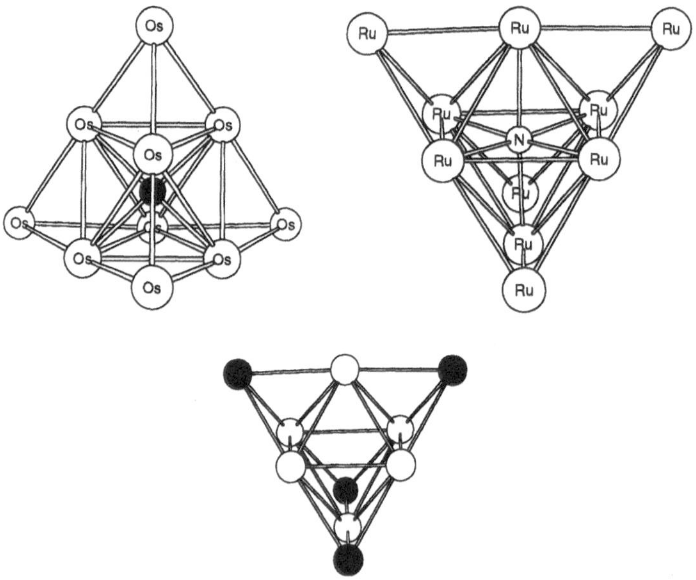

Fig. 6. The carbide and nitride-containing tetra-capped octahedral cores of $[HOs_{10}(CO)_{24}C]^-$ and $[Ru_{10}(CO)_{24}N]^-$, and a representation of the M_{10}-skeleton highlighting the four capping sites

matter how the metal atoms and ligands are grouped. In general, Wade's counting scheme gives no definitive information about the positions (distribution, or whether terminal or bridging) of the CO or other ligands in a cluster. Neither does the classification of $[HOs_{10}(CO)_{24}C]^-$ as a capped *closo*-cage provide information about the positions of the four caps with respect to the eight faces of the octahedron, although consideration of steric effects would tend to suggest an arrangement as is observed in practice (Fig. 6). A similar scheme can be counted for $[Ru_{10}(CO)_{24}N]^-$ as for $[HOs_{10}(CO)_{24}C]^-$; the N atom is isoelectronic with the (C + H) combination, and moving one place to the right in the periodic table means that one fewer cluster bound H atoms are needed. The lower diagram in Fig. 6 shows the M_{10}-skeleton that is common to both $[HOs_{10}(CO)_{24}C]^-$ and $[Ru_{10}(CO)_{24}N]^-$. There are five interstitial cavities, one octahedral and four tetrahedral. The latter are too small to accommodate a first row *p*-block atom and so the electron counting scheme necessarily *suggests* (but cannot confirm) the placement of the carbon or nitrogen atom within the octahedral cavity.

It is important to put into perspective the predictive power of Wade's rules vs specific structural details that single crystal X-ray diffraction studies may reveal. The cluster core of $HRu_6(CO)_{16}(AuPPh_3)_2B$ [17] is shown in Fig. 7. Structural data for this cluster reveal that one Ru-Ru contact is long (339 pm) and outside the 'normal' range of Ru-Ru bonding distances. A comparison of Figs. 7a and 7b might suggest that the Ru_6Au_2-skeleton could be described in terms of a triangular-faced dodecahedron, with the boron atom residing within one part of the cavity. On the other hand, by Wade's approach, the cluster possesses seven electron pairs for bonding the skeletal atoms, and with the knowledge that the gold centres are not considered to be part of the parent cluster skeleton, $HRu_6(CO)_{16}(AuPPh_3)_2B$ fits into the category of a *closo*-octahedral cluster. Another view of the core of the compound is shown in Fig. 7c and here we can see a structural description based upon a distorted octahedral Ru_6-cage with two face-capping $AuPPh_3$ groups. That the distortion to the Ru_6 cage is brought about by the desire for the two gold centres to come within bonding

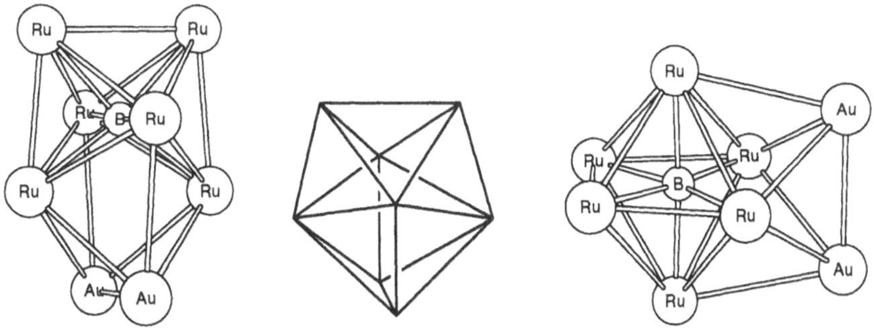

Fig. 7. **a** A view of the core of $HRu_6(CO)_{16}(AuPPh_3)_2B$ which makes it appear to be based upon a dodecahedral skeleton. **b** A dodecahedral skeleton. **c** A second view of the core of $HRu_6(CO)_{16}(AuPPh_3)_2B$ which more clearly illustrates the octahedral core, capped by two associated gold centres

distance (288 pm) appears to be a reasonable explanation for the *apparent* violation of Wade's rules. There *is no violation* – other effects mitigate against the retention of an obviously *closed* octahedral core. Bond lengthening of this type has been observed elsewhere, for example the Ir-Ir interaction in $Ir_2Ru_4(CO)_{16}\{AuP(C_6H_{11})_3\}B$ [22], and illustrates just one of the ambiguities that may arise in structural descriptions.

3 Fragments of the Octahedron: *nido*- and *arachno*-Clusters

Wade's initial premises concerning skeletal electron counting schemes built upon sequences of related compounds, perhaps one of the central of which was (and is when one is teaching Wade's theories) the series $[B_6H_6]^{2-}$, B_5H_9 and B_4H_{10}. As Fig. 8 shows, the arrangements of the boron atoms in each cage are

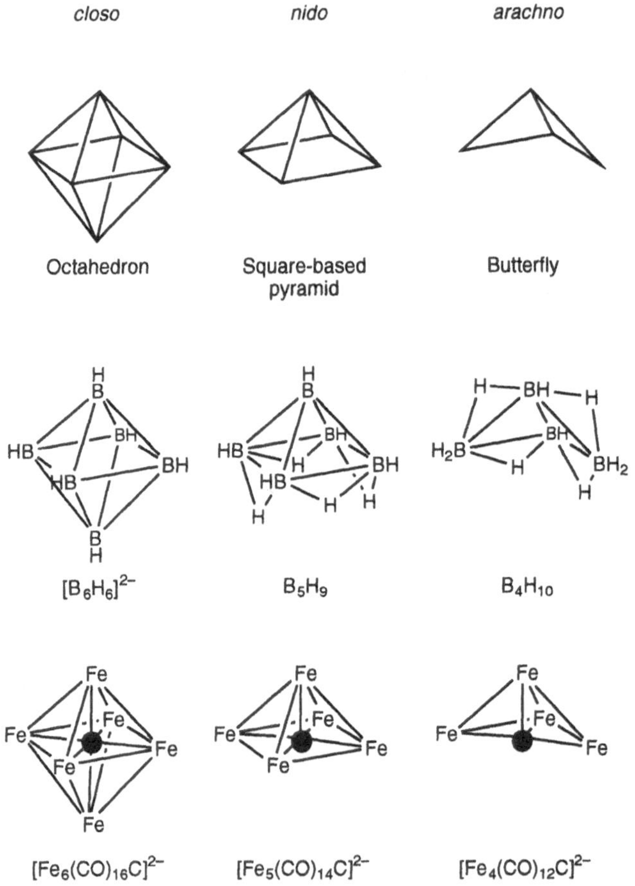

Fig. 8. The relationships between *closo*, *nido* and *arachno* cages in which the *closo*-cage is an octahedron. Examples include the borane and iron carbide cluster series shown

related. In $[B_6H_6]^{2-}$, the boron atoms define a closed (*closo*) octahedron. In B_5H_9 they define a square-based pyramid which is derived from the octahedron by the removal of one vertex. Loss of a vertex from this nest-like cage (*nido*) gives the web-like (*arachno*) arrangement of boron atoms observed in B_4H_{10}. Each borane cluster possesses seven electron pairs and the presence of bridging hydrogen atoms around the open faces of the *nido* and *arachno* cages is typical of such open cages. Isolobal with $[B_6H_6]^{2-}$ is $[Ru_6(CO)_{18}]^{2-}$ but, as we have seen, for electron counting purposes two CO ligands can be 'replaced' by an interstitial carbon atom. Figure 8 shows the structure of $[Fe_6(CO)_{16}C]^{2-}$ [23] which is the first member of a series of *closo*-, *nido*- and *arachno*-species, each successive pair in which differs by one $Fe(CO)_2$ fragment [24, 25]. Since $\{Fe(CO)_2\}$ provides no skeletal electrons (Table 1), the removal of the vertex does not change the skeletal electron count and each member of the series retains seven electron pairs for cluster bonding. Whilst the series in Fig. 8 is a 'paper exercise', similar transformations can be realized in practice and the carbide centre plays an important role in holding the metal framework together during the reactions shown in Eqs. (1) and (2) [26–28].

$$[Fe_6(CO)_{16}C]^{2-} \xrightarrow[\text{}]{\text{Oxidation with Fe(III) or } [C_7H_7]^+} Fe_5(CO)_{15}C \qquad (1)$$

$$Fe_5(CO)_{15}C \xrightarrow{Br^-} [Fe_4(CO)_{12}C]^{2-} \qquad (2)$$

The *nido*- and *arachno*-clusters can be used as building blocks to generate heterometallic *closo*-species. From Fig. 8 and Table 1, Wade's rules would predict that fragments providing zero electrons within their frontier MOs should add to *nido*-$[Fe_5(CO)_{14}C]^{2-}$ to give *closo*-carbide species and this is demonstrated in Eq. (3) [26]. Of course, the reaction is not restricted to the addition of zero-electron units, but if the incoming fragment provides electrons, the cluster may respond by losing an appropriate number of ligands, i.e. it regains the Wadean electron count for the *closo*-cluster. Equation (4) demonstrates that when a Ni(cod) (cod = cycloocta-1,5-diene) unit, equivalent to a $Ni(CO)_2$ unit and a 4-electron donor (Table 1), adds to *nido*-$[Fe_5(CO)_{14}C]^{2-}$, a CO ligand is lost [28]. The product thereby retains seven skeletal electron pairs. We consider later the consequences of simply *adding* electrons without concomitant ligand loss.

$$[Fe_5(CO)_{14}C]^{2-} \xrightarrow{Cr(CO)_3(py)_3} [CrFe_5(CO)_{17}C]^{2-} \qquad (3)$$

$$[Fe_5(CO)_{14}C]^{2-} \xrightarrow{Ni(cod)_2} [Fe_5Ni(CO)_{13}(cod)C]^{2-} \qquad (4)$$

The *arachno*-cluster $[Fe_4(CO)_{12}C]^{2-}$ and its protonated analogues $[HFe_4(CO)_{12}C]^-$ and $[HFe_4(CO)_{12}CH]$ have played a pivotal role in delineating the relationships between cluster-bound organic fragments and those adsorbed on metal surfaces [29, 30]. Isoelectronic with $[Fe_4(CO)_{12}C]^{2-}$ is the boride $[Fe_4(CO)_{12}B]^{3-}$ and it and the protonated derivatives $[Fe_4(CO)_{12}BH]^{2-}$, $[HFe_4(CO)_{12}BH]^-$ and $HFe_4(CO)_{12}BH_2$ have been studied in depth by Fehlner and co-workers [31]. Of these species, the first to be characterized was $HFe_4(CO)_{12}BH_2$. Figure 9 illustrates the then unprecedented environment in which the boron atom resided; the BH_2 unit spanning the wingtips of the Fe_4-butterfly core pro-

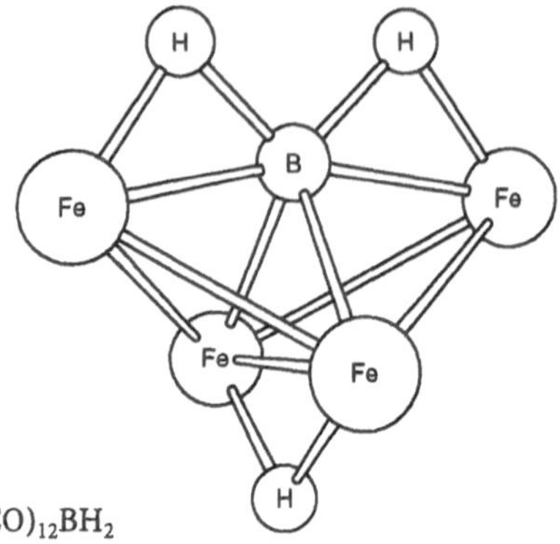

Fig. 9. The core of $HFe_4(CO)_{12}BH_2$

duced problems so far as electron counting was concerned. Should the boron atom be counted as being interstitial or a vertex unit? Careful consideration of the geometrical parameters within the cluster core (Table 2), and the results of a Fenske Hall MO analysis indicated that the boron atom in $HFe_4(CO)_{12}BH_2$ should be regarded as being interstitial, making $HFe_4(CO)_{12}BH_2$ an *arachno*-species [32]. Over later years, more members of this *arachno*-family emerged and several are listed in Table 2; the internal dihedral angles of the M_4-butterflies are similar and are relatively close to the ideal angle of an octahedron (109°) from which the *arachno*-cage is derived.

In $HM_4(CO)_{12}BH_2$ (M = Fe, Ru, Os) [32–24], the boron is prevented from being in an exposed semi-interstitial environment (compare Fig. 9 with Fig. 1) by the presence of the two bridging hydrogen atoms. In order to expose the boron atom, deprotonation is one option, and loss of H^+ does not change the number of skeletal electrons. However, Wade's electron counting rules can be used to suggest possible modifications to the metal butterfly framework that would allow the cluster to attain its quota of cluster-bonding electrons without the need for all three of the cluster-bound H atoms. For example, $HRu_4(CO)_{12}BH_2$ can be prepared from the reaction of $[Ru(CO)_3Cl_2]_2$ with $[Ru_3(CO)_9BH_4]^-$, and similarly, $Ru_4H(\eta^6\text{-}C_6H_6)(CO)_9BH_2$ is formed when $[Ru(\eta^6\text{-}C_6H_6)(MeCN)_3]^{2+}$ reacts with $[Ru_3(CO)_9BH_4]^-$ [37]. In both reactions, a

Table 2. Structural parameters that describe the M_4B framework in related *arachno*-borido-species. The angles should be compared with the ideal angle of 109° within an octahedron from which the *arachno*-M_4 cage is derived

Compound	Internal dihedral angle of the M_4-framework/deg	Ref.
$HFe_4(CO)_{12}BH_2$	114.0	[32]
$HRu_4(CO)_{12}BH_2$	118	[33]
$HOs_4(CO)_{12}BH_2$	113	[34]
$H_2Ru_5(CO)_{13}(\eta^5\text{-}C_5Me_5)BH_2$	114.2	[35]
$HRhRu_4(CO)_{13}(\eta^5\text{-}C_5Me_5)BH_2$	116.6	[36]

2-electron fragment is provided – $Ru(CO)_3$ from $[Ru(CO)_3Cl_2]_2$, and $Ru(\eta^6\text{-}C_6H_6)$ from $[Ru(\eta^6\text{-}C_6H_6)(MeCN)_3]^{2+}$ – and the boron retains two B-H bonding interactions in a core-structure as shown in Fig. 9. If we alter the incoming fragment to a 3-electron donor, then the product might be expected to possess one fewer cluster bound H atoms and the boron atom is one step nearer to being exposed. This is realized in the reaction of $[W(\eta^5\text{-}C_5H_5)(CO)_3]_2$ with $[Ru_3(CO)_9BH_4]^-$; $[W(\eta^5\text{-}C_5H_5)(CO)_3]_2$ is a source of the fragment $W(\eta^5\text{-}C_5H_5)(CO)_2$ and the observed product is $HRu_3W(\eta^5\text{-}C_5H_5)(CO)_{11}BH$ [38]. Each of $HRu_4(CO)_{12}BH_2$, $Ru_4H(\eta^6\text{-}C_6H_6)(CO)_9BH_2$ and $HRu_3W(\eta^5\text{-}C_5H_5)(CO)_{11}BH$ is an *arachno*-species, but the environment of the boron atom can be tuned by varying the electronic properties of the metal fragments.

4
Addition of Electrons: Increasing the Size of the Parent Polyhedron

The conversion of $[Ru_3(CO)_9BH_4]^-$ to $HRu_4(CO)_{12}BH_2$ described above converts a *nido*-cluster based on a trigonal bipyramid into an *arachno*-cluster based on an octahedron. The reaction is of particular interest because it involves the conversion of a vertex BH unit (a 2-electron cluster unit) into a semi-interstitial boron atom plus a cluster-bound hydrogen atom (four cluster electrons in total). More straightforward examples of expansion of the parent-polyhedron are quite common and are exemplified by the reaction of $Ru_5(CO)_{12}(\eta^6\text{-}C_6H_6)C$ with carbon monoxide [39]. Addition of two electrons in the form of the Lewis base CO to *nido*-$Ru_5(CO)_{12}(\eta^6\text{-}C_6H_6)C$ (based on an octahedron) creates *arachno*-$Ru_5(CO)_{13}(\eta^6\text{-}C_6H_6)C$, the Ru_5-core of which is derived from a pentagonal bipyramid (Fig. 10).

A similar example is the addition of an alkyne to a cluster. For example, $PhC\equiv CPh$ adds to the anion *arachno*-$[Ru_4(CO)_{12}N]^-$ in the presence of H^+ to

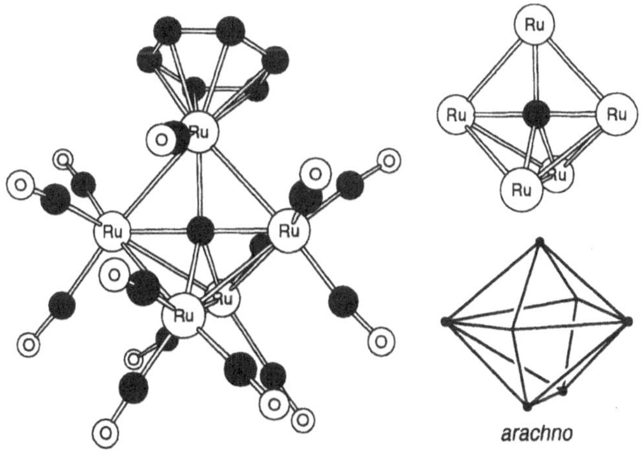

Fig. 10. The structure and core of *arachno*-$Ru_5(CO)_{13}(\eta^6\text{-}C_6H_6)C$, and the relationship of the Ru_5-skeleton to a pentagonal bipyramid

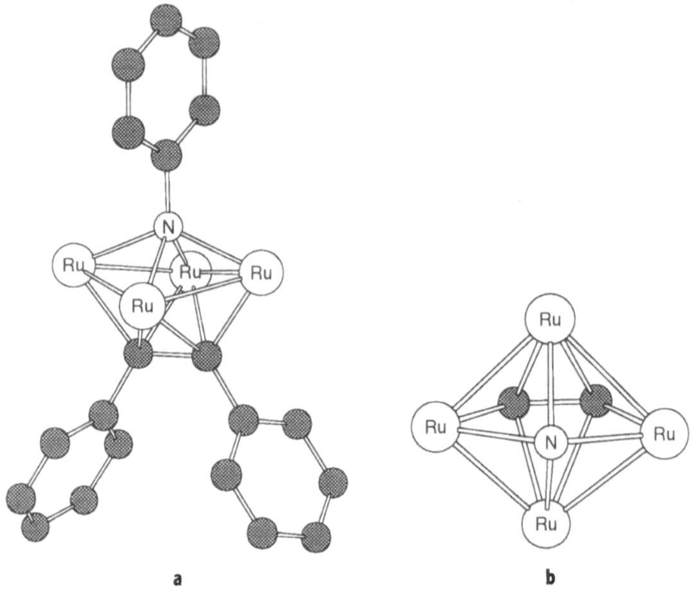

Fig. 11. **a** The structure of $Ru_4(CO)_{12}(PhCCPh)NPh$. **b** The Ru_4NC_2-core defines a pentagonal bipyramid

yield *closo*-$Ru_4(CO)_{12}(PhCCPh)NH$ [40]. Initially, the nitrogen atom (a 5-electron donor) is in a semi-interstitial site within an Ru_4-butterfly framework and the cluster possesses seven pairs of skeletal electrons. During the reaction with the alkyne, the nitrogen atom is protonated and is drawn out of the metal framework to become a vertex atom (a 4-electron NH group); $Ru_4(CO)_{12}(PhCCPh)NH$ is an 8-electron pair *closo*-cage. The same structural type has been observed for $Ru_4(CO)_{12}(PhCCPh)NPh$ (Fig. 11) [41]. In the conversion of *nido*-$Ru_5(CO)_{12}(\eta^6\text{-}C_6H_6)C$ to *arachno*-$Ru_5(CO)_{13}(\eta^6\text{-}C_6H_6)C$ described above, the number of vertex atoms remained the same and the metal core simply adopted a new shape based upon a new parent polyhedron. In contrast, the conversion of *arachno*-$[Ru_4(CO)_{12}N]^-$ to *closo*-$Ru_4(CO)_{12}(PhCCPh)NH$ involves (i) the addition of four electrons from the alkyne and (ii) the transformation of a (formally) 6-electron interstitial N^- unit into a 4-electron vertex NH unit. The net gain in cluster bonding electrons is thus two and the parent polyhedron changes from an octahedron to a pentagonal bipyramid (Fig. 11).

5
The Trigonal Prism: A Problematical Geometry?

The clusters $[Co_6(CO)_{13}C]^{2-}$ [42] and $[Co_6(CO)_{13}N]^-$ (Fig. 12a) [43] are both *closo*-octahedral cages with interstitial carbon atoms. In contrast, $[Co_6(CO)_{15}C]^{2-}$ (Fig. 12b) [44] and $[Co_6(CO)_{15}N]^-$ [45] possess trigonal prismatic metal frames. Each of $[Co_6(CO)_{15}C]^{2-}$ and $[Co_6(CO)_{15}N]^-$ possesses nine skeletal electron pairs

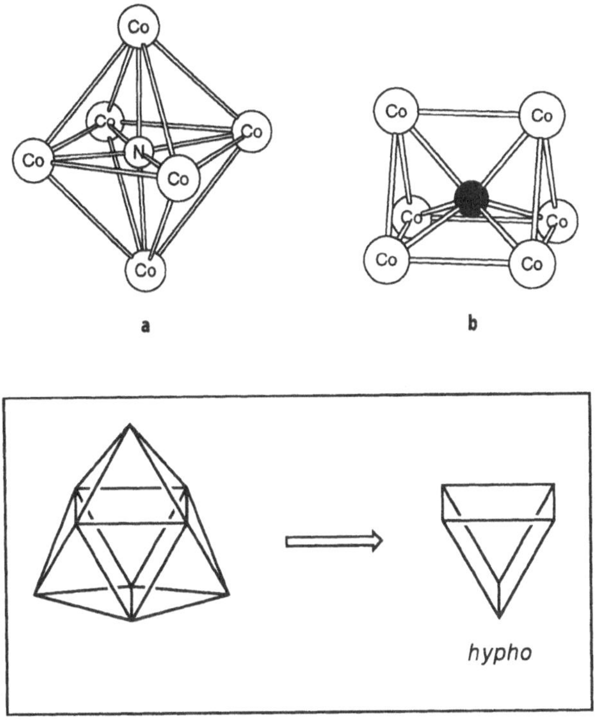

c

Fig. 12 a–c. The cores of: **a** $[Co_6(CO)_{13}N]^-$; **b** $[Co_6(CO)_{15}C]^{2-}$; **c** Wade's approach would consider a trigonal prismatic species to be a *hypho*-cluster based upon a tricapped trigonal prism

and might be expected therefore to be *arachno* clusters based upon the eight vertex dodecahedron. The observed trigonal prismatic framework may be considered by Wade's approach to be a *hypho*-cluster based upon a tricapped trigonal prism but this description would require ten electron pairs for cluster bonding. Do we therefore have a problem – a violation of the rules?

The trigonal prism is a rather special case – in fact, there is no trigonal prismatic borane prototype on which to base an isolobal cluster analogue. One of the simplest examples of a trigonal prismatic cluster is prismane C_6H_6 in which the nine bonding pairs of electrons available from six CH units are accommodated in nine *localized* C-C single bonds. Likewise, one may consider a trigonal prismatic *d*-block metal carbonyl cluster to possess nine localized M-M bonds but in the large majority of trigonal prismatic examples, the interstice is occupied. In a cluster such as $[Co_6(CO)_{15}C]^{2-}$ we may still think in terms of nine pairs of electrons being compatible with nine M-M edges; the valence *s* and *p* orbitals of the *p*-block interstitial atom overlap strongly with the skeletal MOs of the M_6-cage and contribute all their valence electrons to cage bonding [46]. Of course, this means that the notion of localized M-M bonds is not a wholly satisfactory picture because the role of the interstitial atom is crucial.

This leads us away from the fundamental rules of Wade (namely a parent *deltahedral* – triangular-faced – skeleton with n vertices requiring $(n+1)$ electron pairs) to the more generally applicable Polyhedral Skeletal Electron Pair Theory (PSEPT) [47]. Within its remit, this approach allows for a greater variety of polyhedral types than the deltahedra focused upon by Wade. A trigonal prism is classed as a 'three-connected polyhedron' and by PSEPT such a *metal* polyhedron with m vertices requires a total of $15m$ valence electrons. Thus, both $[Co_6(CO)_{15}C]^{2-}$ and $[Co_6(CO)_{15}N]^-$ are 90-electron trigonal prismatic clusters as are $[H_2Ru_6(CO)_{18}B]^-$ [17] and $[Os_6(CO)_{18}P]^-$ [48].

So why does the six-vertex octahedron 'obey' Wade and the six-vertex trigonal prism 'violate' Wade? We have already alluded to the crucial point – the octahedron is a deltahedron and has a prototype borane cluster in $[B_6H_6]^{2-}$. Closed *deltahedral* clusters with n vertex atoms in high connectivity sites approximate to spheres and, for each, there are $(n + 1)$ skeletally-bonding molecular orbitals. Removing a vertex does not alter the number of cage-bonding MOs and so we can build up sets of related clusters analogous to the *closo*-, *nido*-, *arachno*- and *hypho*- (etc.) skeletons. Within the PSEPT approach, a closed m vertex *metal deltahedron* requires $(14m + 2)$ valence electrons, whilst the *nido*- and *arachno* derivatives require a total of $(14m + 4)$ and $(14m + 6)$ valence bonding electrons respectively. Thus, $Ru_6(CO)_{17}C$, which, as we saw earlier, is classed as a 7-electron *closo*-cluster by Wade, is also an 86-electron cluster by PSEPT as are other octahedral metal clusters.

6 Bicapped Antiprismatic Clusters and *nido*- and *arachno*-Derivatives

The anion $[B_{10}H_{10}]^{2-}$ possesses a bicapped square-antiprismatic structure and is a *closo*-species with 11 pairs of cluster bonding electrons. The same skeletal shape is observed in, for example, $[Rh_{10}(CO)_{22}As]^{3-}$ [49] (Fig. 13a) and the skeletal electron count for this anion is also consistent with a *closo*-cage. Interstitial arsenic and antimony atoms are not well exemplified: this must reflect the greater steric requirements of these atoms compared to their first

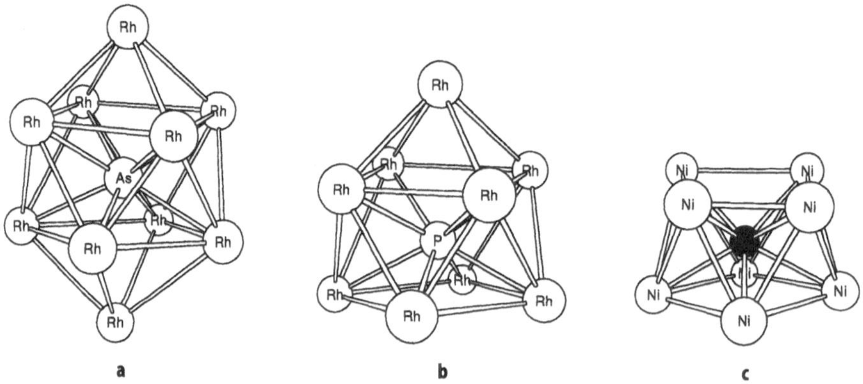

Fig. 13a–c. The cores of: **a** $[Rh_{10}(CO)_{22}As]^{3-}$; **b** $[Rh_9(CO)_{21}P]^{2-}$; **c** $[Ni_8(CO)_{16}C]^{2-}$

and second row congeners and the need for high nuclearity metal clusters with relatively open structures; many high nuclearity metal clusters consist of condensed structures with predominantly tetrahedral, octahedral and prismatic interstices. The cavity within a *square-antiprism* can accommodate a third row *p*-block atom; inspection of the Rh-As bond distances in $[Rh_{10}(CO)_{22}As]^{3-}$ shows that the arsenic atom has primary interactions with the metal atoms of the central Rh_8-cavity (av $Rh_{antiprism}$-As 249 pm; av Rh_{apex}-As 300 pm). For accommodation of a fourth row atom (e.g. antimony) we must go to the larger icosahedral cage and $[Rh_{12}(CO)_{27}Sb]^{3-}$ [51] is a rare example of such a *closo*-species possessing 13 cluster bonding pairs analogous to the borane prototype $[B_{12}H_{12}]^{2-}$.

Removal of a vertex from the bicapped square antiprism without a change in the skeletal electron count is observed when the phosphide $[Rh_{10}(CO)_{22}P]^{3-}$ reacts with CO to give *nido*-$[Rh_9(CO)_{21}P]^{2-}$ [50] (Fig. 13b); the reaction is reversed by treatment of the *nido*-cluster anion with $[Rh(CO)_4]^-$. This 10 $\rightleftharpoons$ 9-vertex structural transformation (although rarely exemplified in practice) is exactly as predicted by Wade. In theory, a second vertex atom could be removed to give an *arachno*-cage in which the square-antiprismatic cavity was still intact. In practice, the decapping stops at $[Rh_9(CO)_{21}P]^{2-}$, although other examples of square-antiprismatic M_8-clusters containing interstitial atoms are known. One such is $[Ni_8(CO)_{16}C]^{2-}$ [52] (Fig. 13c) which possesses 11 pairs of cluster bonding electrons and is thus classed as an *arachno*-species by Wade. The group of clusters structurally similar to $[Ni_8(CO)_{16}C]^{2-}$ is rather small, but analysis of the number of cluster bonding electrons in each delineates the group into two sub-classes. Whereas $[Ni_8(CO)_{16}C]^{2-}$ is a 'well-behaved' Wadean *arachno*-cluster, $[Co_8(CO)_{18}C]^{2-}$ [53] and $Rh_8(CO)_{19}C$ [54] possess only nine pairs of skeletal electrons. This problem is addressed by classifying the square-antiprism as a 'four-connected polyhedron': within PSEPT, a four-connected metal polyhedron with *m* vertices requires a total number of $(14m + 2)$ valence electrons. Both $[Co_8(CO)_{18}C]^{2-}$ and $Rh_8(CO)_{19}C$ fit this criterion, each having 114 valence electrons. PSEPT recognizes that a square-antiprismatic metal cluster may exist with either 114 valence electrons (a four-connected classification) or with 118 valence electrons (the Wade *arachno* classification, written in terms of total valence, rather than skeletally-bonding, electrons) and the electronic flexibility is manifested in the observation that $[Co_8(CO)_{18}C]^{2-}$ undergoes three electrochemically reversible 1-electron processes [55].

7
Condensed Polyhedra

Wade's fundamental concepts are generally successful for tetrahedral, trigonal bipyramidal and octahedral clusters and capped derivatives of these species. Of these, only the octahedral cavity may accommodate a *p*-block element and even then we are restricted on steric grounds to boron, carbon, nitrogen and (very rarely in low oxidation state metal clusters [56]) oxygen atoms. A cluster such as $[Ru_{10}(CO)_{24}N]^-$ [21] fits nicely into the category of a tetracapped *closo* species, but it is also usefully considered as a *condensed polyhedral* assembly consisting of an octahedron condensed with four tetrahedra (Fig. 6).

The condensation of polyhedra is a basic method of cluster growth and, in many cases, mimics the assembly of a section of an array of metal atoms in the bulk state. If we restrict the discussion to condensed clusters containing interstitial atoms, then the capped octahedra represent just one sub-group. Figure 14 provides four examples of structurally characterized clusters which are *not* based upon the Wadean capping principle. The anion $[HFe_7(CO)_{20}B]^{2-}$ [57] (Fig. 14a) consists of a face-sharing trigonal prism and square-based pyramid; the boron atom resides within the trigonal prismatic cage and is rather remote from the capping iron atom (Fe-B distances = 236 pm vs. 202–214 pm). In contrast, in the structurally related $[Ni_7(CO)_{12}C]^{2-}$ [58], the interstitial carbon atom is displaced from the centre of the prismatic cavity towards the capping nickel atom. It is not easy to rationalize these metal frameworks in terms of Wade; they may appear to be *arachno*-clusters based upon the tricapped trigonal prism, but neither possesses the necessary ten pairs of cluster bonding electrons. With the remit of PSEPT however, the total valence electron count for a condensed cluster is equal to the sum of the electron counts for the component polyhedra (Table 3) *minus* the number of electrons associated with the shared face of metal atoms, the shared vertex atoms or the shared pairs of atoms (i.e. an edge, see below). Condensing a trigonal prism and square-based pyramid through a square-face leads to a total valence electron count of 100 or 102 electrons (Table 3). The anion $[Ni_7(CO)_{12}C]^{2-}$ satisfies the former electron count whilst $[HFe_7(CO)_{20}B]^{2-}$ satisfies the latter.

Figure 14b illustrates $[Co_6Ni_2(CO)_{16}C_2]^{2-}$, an example of a face-sharing double trigonal prism, with each interstitial site containing a carbon atom [59]. A related cluster is $[Rh_3Ru_6(CO)_{23}B_2]^-$ – in which a Rh_2Ru_6-double prism is additionally capped by a rhodium vertex [60]. These two clusters form part of a series of related systems [61]. Such condensed clusters cannot be viewed with the remit of Wade's rules, but are successfully treated by PSEPT: $[Co_6Ni_2(CO)_{16}C_2]^{2-}$ is a 116-electron cluster and $[Rh_3Ru_6(CO)_{23}B_2]^-$ possesses 128 valence electrons.

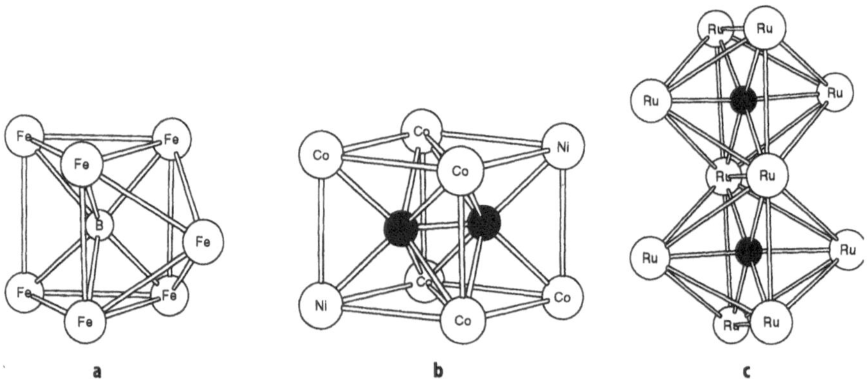

Fig. 14a–c. The cores of: **a** $[HFe_7(CO)_{20}B]^{2-}$; **b** $[Co_6Ni_2(CO)_{16}C_2]^{2-}$; **c** $[Ru_{10}(CO)_{24}C_2]^{2-}$

Table 3. Characteristic total valence electron counts for selected *d*-block metal cages and electron counts associated with shared units of metal atoms

Cluster cage geometry	Valence electron count	Shared units of metal atoms	Valence electron count
Triangle	48	Common vertex	18
Tetrahedron	60	Common edge	34
Butterfly or a planar, 4-atom raft	62	Common triangular face	48
		Common square face	62 or 64
Square	64		
Trigonal bipyramid	72		
Square-based pyramid	74		
Octahedron	86		
Trigonal prism	90		

The anion $[Ru_{10}(CO)_{24}C_2]^{2-}$ [62] (Fig. 14c) exemplifies condensation by edge-sharing. Once again, consideration of the total valence electron count is the most satisfactory way in which to approach a rationalization of this structure. Fusing two octahedra through an M-M edge gives a skeleton that requires 138 valence electrons (Table 3), the number possessed by $[Ru_{10}(CO)_{24}C_2]^{2-}$.

8 Clusters with Gold(I) Phosphine Units Surrounding a *p*-Block Atom

Over the last few years, the work of Schmidbauer has illustrated a new type of interstitial environment for some of the *p*-block elements. Although gold(I) fragments have been used extensively as isolobal hydrogen-replacements in metal carbonyl clusters (see Sect. 2), their involvement as cluster *vertex* units to give a cage encapsulating a non-metal atom has been less well exemplified, although the area is now being exploited with intriguing results. Figure 15 shows three examples: $[(Ph_3P)_4Au_4N]^+$ [63], $[(Ph_3P)_5Au_5N]^{2+}$ [64] and $[(Ph_3P)_6Au_5P]^{2+}$ [65]. In $[(Ph_3P)_4Au_4N]^+$, (which may be formally considered to be an analogue

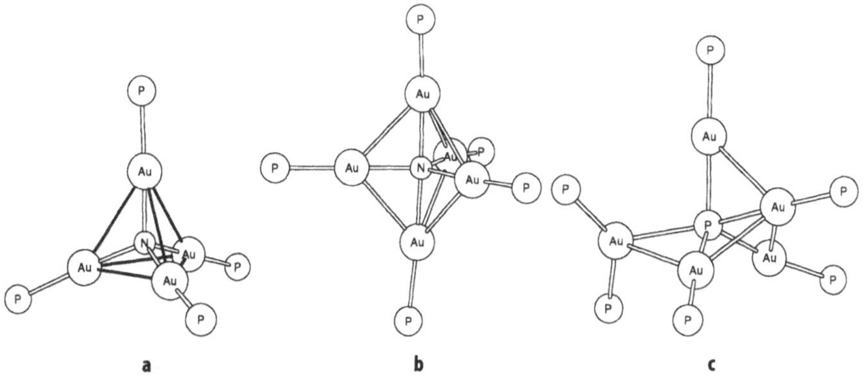

Fig. 15a–c. The cores of cations: **a** $[(Ph_3P)_4Au_4N]^+$; **b** $[(Ph_3P)_5Au_5N]^{2+}$; **c** $[(Ph_3P)_6Au_5P]^{2+}$

of the ammonium ion in which H atoms have been replaced by $AuPPh_3$ groups), the Au-Au bond distances of 334.5 and 322.7 pm are rather long to be termed bonding interactions. The importance of the radial component in the bonding interactions of gold-containing fragments is well established and it is twenty years since Mingos pointed out that the valence orbitals of gold atoms do not generate a strongly bonding set of molecular orbitals tangential to a polyhedral surface in an Au_x skeleton [66]. Neither Wade's rules nor PSEPT is applicable to clusters such as $[(Ph_3P)_4Au_4N]^+$. In the cation $[(Ph_3P)_5Au_5N]^{2+}$, the Au-Au distances fall into two sets: the apex-equatorial distances (Au-Au range 288 to 307 pm) are consistent with bonding interactions and are indicated in Fig. 15b by Au-Au connections while the equatorial Au-Au distances are significantly longer (339–373 pm). Again, radial Au-N interactions are a crucial component in stabilizing the species. In $[(Ph_3P)_6Au_5P]^{2+}$, the Au atoms define a rather obscure polyhedral framework around the central phosphorus atom, and the role of the radial Au-P bonding is obvious.

9 Conclusions and Comments

Deltahedral anions in the family $[B_nH_n]^{2-}$ are the *closo*-clusters from which a wide variety of other boranes and heteroboranes are structurally derived; Wade's rules were developed for such species and are remarkably successful both in rationalizing structures and predicting them. The $[B_nH_n]^{2-}$ anions provide a series of prototypes on which families of metal clusters can be based by using isolobal-fragment replacements. Steric interactions between ligands may mitigate against the formation of a particular molecule or molecular ion but the placement of an atom within the interstice of a cluster with n *3* 6 produces a stabilizing effect; the interstitial atom provides all of its valence electrons to cluster bonding without contributing to *exo*-ligand steric effects. The application of Wade's rules to such *deltahedral* species is generally successful, although ambiguities may arise, for example when M-M contacts are apparently rather long to be considered 'bonding'.

Whereas the boranes show a preference for deltahedral cages, the options for polyhedral frameworks of metal atoms are not as restricted. For example, the trigonal prism (which may accommodate an interstitial atom from either the first or second row of the *p*-block) is a common motif in cluster chemistry. Rationalization of the bonding in these non-deltahedral clusters is best approached using PSEPT. Some geometries are ambiguous in their origins; the square-antiprism may be considered to be an *arachno*-Wadean cluster or a 4-connected PSEPT-cluster, and both designations are shown to be realistic.

Wade's rules are typically *not* appropriate for condensed clusters, unless the framework fits into the category of a capped deltahedron. Again, a consideration of the total valence electrons count (i.e. PSEPT) proves to be more generally useful.

10 References

1. Wade K (1976) Adv Inorg Chem Radiochem 18:1
2. Wade K (1975) Chem Brit 11:177
3. Mingos DMP, Wales DJ (1990) Introduction to cluster chemistry. Prentice Hall, New Jersey
4. Housecroft CE (1996) Metal-metal bonded carbonyl dimers and clusters. Oxford University Press, Oxford
5. O'Neill ME, Wade K (1982) Structural and bonding relationships among main group organometallic compounds. In: Wilkinson G, Stone FGA, Abel E (eds) Comprehensive organometallic chemistry. Pergamon, Oxford, vol 1, p 1
6. Housecroft CE (1994) Cluster molecules of the *p*-block elements. Oxford University Press, Oxford
7. Mingos DMP, May AS (1990) Structural and bonding aspects of metal carbonyl cluster chemistry. In: Shriver DF, Kaesz HD, Adams RD (eds) The chemistry of metal cluster complexes. VCH, New York, p 11
8. Braga D, Grepioni F, Dyson PJ, Johnson BFG, Frediani P, Bianchi M, Piacenti F (1992) J Chem Soc Dalton Trans 2565
9. Jackson PF, Johnson BFG, Lewis J, McPartlin M, Nelson WJH (1979) J Chem Soc Chem Comm 735
10. Hsu G, Wilson SR, Shapley JR (1991) Inorg Chem 30:3881
11. Albano VG, Braga D, Martinengo S (1986) J Chem Soc Dalton Trans 981
12. Farrugia LJ (1988) Acta Crystallogr Sect C 44:997
13. Dyson PJ, Johnson BFG, Lewis J, Martinelli M, Braga D, Grepioni F (1993) J Am Chem Soc 115:9062
14. Beringhelli T, D'Alfonso G, de Angelis M, Ciani G, Sironi A (1987) J Organomet Chem 322:C21
15. Ciani G, D'Alfonso G, Freni M, Romiti P, Sironi A (1982) J Chem Soc Chem Comm 339
16. Henly TJ, Shapley JR, Rheingold AL (1986) J Organomet Chem 310:55
17. Housecroft CE, Matthews DM, Waller A, Edwards AJ, Rheingold AL (1993) J Chem Soc Dalton Trans 3059
18. Lauher JW, Wald K (1981) J Am Chem Soc 100:5305
19. Hall KP, Mingos DMP (1984) Prog Inorg Chem 32:237
20. Jackson PF, Johnson BFG, Lewis J, McPartlin M, Nelson WJH (1982) J Chem Soc Chem Comm 49
21. Bailey PJ, Conole G, Johnson BFG, Lewis J, McPartlin M, Moule A, Powell HR, Wilkinson DA (1995) J Chem Soc Dalton Trans 741
22. Galsworthy JR, Hattersley AD, Housecroft CE, Rheingold AL, Waller, A (1995) J Chem Soc Dalton Trans 549
23. Churchill MR, Wormwald J (1974) J Chem Soc Dalton Trans 2410
24. Gourdon A, Jeannin Y (1985) J Organomet Chem 290:199
25. Boehme RF, Coppens P (1981) Acta Crystallogr Sect B 37:1914
26. Tachikawa T, Sievert AC, Muetterties EL, Thompson MR, Day CS, Day VW (1980) J Am Chem Soc 102:1725
27. Bradley JS, Hill EW, Ansell GB, Modrick MA (1982) Organometallics 1:1634
28. Braye EH, Dahl LF, Hubel W, Wampler DL (1962) J Am Chem Soc 84:4633
29. Bradley JS, (1983) Adv Organometal Chem 22:1 and references cited therein.
30. Drezdon MA, Whitmire KH, Bhattacharyya AA, Hsu W-L, Nagel CC, Shore SG, Shriver DF (1982) J Am Chem Soc 104:5630
31. Housecroft CE (1995) Coord Chem Rev 143: 297 and references cited therein
32. Housecroft CE, Fehlner TP, Scheidt WR, Wong KS (1983) Organometallics 2:825
33. Hong F-E, McCarthy DA, White JP, Cottrell CE, Shore SG, (1990) Inorg Chem 29:2874
34. Chung J-J, Knoeppel D, McCarthy D, Columbie A, Shore SG (1993) Inorg Chem 32:3391

35. Galsworthy JR, Housecroft CE, Rheingold AL (1993) 4167
36. Galsworthy JR, Housecroft CE, Edwards AJ, Raithby PR (1995) J Chem Soc Dalton Trans 2935
37. Galsworthy JR, Housecroft CE, Nixon DM, Rheingold AL (1996) J Organomet Chem (in press)
38. Housecroft CE, Matthews DM, Rheingold AL, Song X (1992) J Chem Soc Dalton Trans 2855
39. Braga D, Grepioni F, Sabatino P, Dyson PJ, Johnson BFG, Lewis J, Bailey PJ, Raithby PR, Stalke D (1993) J Chem Soc Dalton Trans 985
40. Blohm M, Gladfelter WL (1986) Organometallics 5:1049
41. Rheingold AL, Staley DL, Han S-H, Geoffroy GL (1988) Acta Crystallogr Sect C 44:570
42. Albano VG, Braga D, Martinengo S (1985) Eur Cryst Meeting 9:162
43. Ciani G, Martinengo S (1986) J Organomet Chem 306:C49
44. Martinengo S, Strumolo D, Chini P, Albano VG, Braga D (1985) J Chem Soc Dalton Trans 35
45. Martinengo S, Ciani G, Sironi A , Heaton BT, Mason J (1979) J Am Chem Soc 101:7095
46. See for example: Wijeyesekera SD, Hoffmann R (1984) Organometallics 3:949
47. Mingos DMP, May AS (1990) Structural and bonding aspects of metal carbonyl cluster chemistry. In: Shriver DF, Kaesz HD, Adams RD (eds) The chemistry of metal cluster complexes. VCH, New York and references cited therein
48. Colbran SB, Lahoz FJ, Raithby PR, Lewis J, Johnson BFG, Cardin CJ (1988) J Chem Soc Dalton Trans 173
49. Vidal JL (1981) Inorg Chem 20:243
50. Vidal JL, Walker WE, Pruett RL, Schoening RC (1979) Inorg Chem 18:129
51. Vidal JL,Troup JM (1981) J Organomet Chem (1981) 213:351
52. Ceriotti A, Longoni G, Manassero M, Perego M, Sansoni M (1985) Inorg Chem 24:117
53. Albano VG, Chini P, Ciani G, Martinengo S, Sansoni M (1978) J Chem Soc Dalton Trans 463
54. Albano VG, Sansoni M, Chini P, Martinengo S, Strumolo D (1975) J Chem Soc Dalton Trans 305
55. Rimmelin J, Lemoine P, Gross M, Mathieu R, De Montauzon D (1986) 309:355
56. Schauer CK, Shriver DF (1987) Angew Chem Int Ed Engl 26:255
57. Bandyopadhyay A, Shang M, Jun C-S, Fehlner TP (1994) Inorg Chem 33:3677
58. Ceriotti A, Piro G, Longoni G, Manassero M, Masciocchi N, Sansoni M (1988) New J Chem 12:501
59. Ceriotti A, Della Pergola R, Garlaschelli L, Longoni G, Manassero M, Masciocchi N, Sansoni M, Zanello P (1992) Gazz Chim Ital 122:365
60. Galsworthy JR, Housecroft CE, Rheingold AL (1994) J Chem Soc Dalton Trans 2359
61. Galsworthy JR, Housecroft CE, Rheingold AL (1994) J Chem Soc Dalton Trans 2359 and references therein
62. Hayward CMT, Shapley JR, Churchill MR, Bueno C, Rheingold AL (1982) J Am Chem Soc 104:7347
63. Zeller E, Beruda H, Kolb A, Bissinger P, Riede J, Schmidbaur H (1991) Nature (London) 352:141
64. Grohmann A, Riede J, Schmidbaur H (1990) Nature (London), 345:140
65. Beruda H, Zeller E, Schmidbaur H (1993) Chem Ber 126:2037
66. Mingos DMP (1976) J Chem Soc Dalton Trans 1163.

Diverse Naked Clusters of the Heavy Main-Group Elements. Electronic Regularities and Analogies

John D. Corbett

Department of Chemistry and Ames Laboratory, Iowa State University, Ames, Iowa 50011, *E-mail: JDC@AmesLab.gov*

Wade's rules provide valuable guidelines regarding geometric-electronic interrelations for over two dozen types of polyhedra among (1) Zintl ions of the group 14 and 15 elements isolated from nonaqueous solvents and (2) numerous anionic clusters of particularly In and Tl found in "neat" alkali-hypoelectronic, open-shell, and centered clusters and in networks are also described.

Keywords: metal polyhedra; hypoelectronic (clusters); naked clusters; electronic structure; synthesis; icosahedra; heavy metal clusters; centered clusters.

1 Background

The possibility that classical electron counting rules derived for boron-based deltahedra – often known as Wade's rules [1,2] – would later be found applicable and useful for a large group of homopolyatomic cations and anions of elements well removed from boron was not anticipated; rather the application followed on the syntheses of new examples [3]. That these would be almost exclusively "naked" clusters free of ligands or exo substituents like hydrogen or alkyl groups was perhaps not obvious either, although the increased weakness of these types of sigma bonds is a recognized property of the heavier main-group elements. The following will describe the development of this area of heavier p-element cluster chemistry.

Some hints of these cluster prospects were to be found in the older literature, as is often the case. First, starting in 1953 with NaPb [4], the 1:1 ATt compounds between A = Na, K, Rb, Cs and Tt (tetrel) = Si, Ge, Sn, Pb were shown to contain more or less regular tetrahedra Tt_4 [5], so that these compounds could be described as Zintl phases $(A^+)_4Tt_4^{4-}$ with closed-shell ions [6,7]. (Oxidation states not actual charges should be emphasized in such descriptions.) However, these anions were generally categorized instead as reasonable analogues of the molecules P_4, As_4, etc. rather than nido-deltahedra with $2n+4$ skeletal electrons, which they also are.

A second hint turned out to be misleading. A long-known "bismuth monochloride" was established in the early 1960s to actually have a Bi_6Cl_7 composition and to contain Bi_9^{5+} clusters together with chlorobismuthate(III) anions in a fairly complex structure $[(Bi_9^{5+})_2(BiCl_5^{2-})_4Bi_2Cl_8^{2-}]$ [8]. The bismuth polyhedra were somewhat distorted from the regular tricapped trigonal prismatic configuration but were closest to this limit. (Later work showed that the distortion is not general.) Subsequent extended-Hückel calculations demonstrated that the D_{3h} ideal was closed shell for $Bi_9{}^{5+}$ [9]. However, the classical geometry–electron counting guides for polyhedra were seemingly violated since the 22 skeletal *p* electrons $(9 \cdot 3 - 5)$ exceeded by two the predicted $2n+2$ limit for such a closo-polyhedron. (A *p*-orbital-only basis set seemed reasonable for these calculations in view of the 9–10 eV separation between 6*s* and 6*p* states in Bi^+. This description will be utilized throughout this article, with further information later.) It was also noted [10] that Bi_9^{5+} was isoelectronic, and thence presumably isosteric, with the Pb_9^{4-} and Sn_9^{4-} ions that had been deduced by Zintl in the 1930s from electrochemical and extraction studies in NH_3(l) but never isolated.

Fortunately for the development of the field, two more polycations of bismuth discovered through early applications of acid-stabilization concepts [11] made the suitability of Wade's rules for these types of clusters more evident [12]. The presumed cations in the new solid phases $Bi_5(AlCl_4)_3$ and $Bi_8(AlCl_4)_2$ (Bi_4AlCl_4 empirically) were thought to be closely analogous in composition and spectra to the Bi_5^{3+} and Bi_8^{2+} solutes that had been well characterized in chloroaluminate melts [13,14]. Lacking workable single crystals at the time, the configuration of the cluster in each phase was predicted from EHMO calculations. A trigonal bipyramidal (D_{3h}) Bi_5^{3+} was, not surprisingly, readily distinguished from C_{4v} or T_d prospects. Even a spectral assignment appeared possible on this basis. Likewise, an archimedean antiprismatic configuration (D_{4d}) was predicted for the 22-e Bi_8^{2+}, clearly distinguishable from the cube or the triangular-faced dodecahedron (D_{2d}) which were closed shell only with the implausible 24 or 20 electrons, respectively. The correctness of these assignments was confirmed much later by crystallography [15,16]. At the time it was also noted that the Bi_3^+ ion with a planar D_{3h} configuration would be a suitable model for the solute $Bi_3^+ \cdot yBi^{3+}$ indicated by spectroscopic, emf, and vapor pressure measurements on solutions of up to ~6 mol% Bi in liquid $BiCl_3$. Although no compound of Bi_3^+ has ever been isolated, the diamagnetic Bi^+ ion that forms first in more dilute solutions in $BiCl_3$ or $NaAlCl_4$ was, with luck, later trapped in $Bi^+Bi_9^{5+}(HfCl_6^{2-})_3$ [17].

2 Homoatomic Clusters Based on Classical Bonding Rules

2.1 Overview

It became apparent from the foregoing discoveries and calculations that a general relationship with well-known polyborane(2–) ions or their carboborane relatives could be derived [12]. This was first illustrated for a series of closo bismuth polyhedra by the imaginary conversions:

$$B_nH_n^{2-} \rightarrow (:B)_n^{-n-2} + nH^+ \tag{1}$$

and

$$(:B)_n^{-n-2} \xrightarrow{2p} (:Bi)_n^{n-2} \tag{2}$$

The first step leaves n electron pairs on the n-atom cluster, nonbonding s-pairs in the extreme, in place of those formerly in the exo bonds to hydrogen in the borane. The second step imagines the introduction of two protons into each nucleus to reduce the high cluster charge, thus changing the cluster element into a pnictogen, specifically bismuth. Of course, there's also a change of four periods in this process! Nonetheless, general comparisons of stoichiometries and MO's for several of these quite different isosteric pairs, boranes vs naked heavy-atom clusters, show close parallels in the electron distributions and eigenvalues for skeletal bonding, as might be expected the basis of largely

Table 1. Homoatomic cluster possibilities with classical closed shell configurations for Triel (Tr), Tetrel (Tt) and Pnictogen (Pn) families

	Tr (Al,Ga,In,Tl)	Tt (Si–Pb)	Pn (P–Bi)
closo	Tr_n^{-n-2}	Tt_n^{-2}	Pn_n^{n-2}
nido	Tr_n^{-n-4}	Tt_n^{-4}	Pn_n^{n-4}
arachno	Tr_n^{-n-6}	Tt_n^{-6}	Pn_n^{n-6}

Table 2. Zintl ions and analogues with classical skeletal electron counts[a]

Ion	Compound	Ion Symmetry	Skeletal Electrons		Ref.
Tt_4^{4-}	(Na–Cs)(Si–Pb), $BaSi_2$	T_d (nido)	12	$2n+4$	6
$Sn_2Bi_2^{2-}$	$(K\text{-cp})_2Sn_2Bi_2$	~T_d^b (nido)	12	$2n+4$	19
$Pb_2Sb_2^{2-}$	$(K\text{-cp})_2Pb_2Sb_2$	(disordered)			20
In_4^{8-}	Na_2In	~T_d (nido)	12	$2n+4$	21
Tl_4^{8-}	Na_2Tl				22
Se_4^{2+}, Te_4^{2+}	$Ch_4(AlCl_4)_2$	D_{4h} (arachno)	14	$2n+6$	23, 24
Sb_4^{2-}, Bi_4^{2-}	$(K\text{-cp})_2Pn_4$				25, 26
Bi_5^{3+}	$Bi_5(AlCl_4)_3$	D_h(closo)	12	$2n+2$	12, 16
Sn_5^{2-}, Pb_5^{2-}	$(Na\text{-cp})_2Tt_5$				27
Tl_5^{7-}	$Na_2K_{21}Tl_{19}$	~D_{3h} (closo)	12	$2n+2$	28
	$Na_{23}K_9Tl_{15.3}$	D_{3h} (closo)			29
In_5^{9-}	La_3In_5	C_{4v} (nido)	14	$2n+4$	30
Ga_6^{8-}	$Ba_5Ga_6H_2$	~O_h	14	$2n+2$	31, 32
Tl_6^{8-}	$Na_{14}K_6Tl_{18}M$	O_h			33
Bi_8^{2+}	Bi_4AlCl_4	~D_{4d} (arachno)	22	$2n+4$	12, 15
Ge_9^{2-}	$(K\text{-cp})_6Ge_{18}$	~C_{2v} (closo)	29	$2n+2$	34
Ge_9^{4-}	$(K\text{-cp})_6Ge_{18}$				34
Sn_9^{4-}	$(Na\text{-cp})_4Sn_9$	C_{4v} (nido)	22	$2n+4$	35
	$(K\text{-cp})_3KSn_9$				36
Pb_9^{4-}	$(K\text{-cp})_3KPb_9$				37
$TlSn_8^{3-}$	$(K\text{-cp})_3(TlSn_{17/2})$	~D_{3h}[b] (closo)	20	$2n+2$	38
$TlSn_9^{3-}$		~D_{4d}[b] (closo)	22	$2n+2$	
Ge_{10}^{2-}	$(K\text{-cp})_2Ge_{10}$	~D_{4d}(closo)	22	$2n+2$	39

[a] Ch = chalcogen; Pn = pnictogen (As–Bi); Tt = tetrel (Si–Pb); cp = 2,2,2-crypt.
[b] Symmetry neglecting heteroatoms in cluster shell.

on symmetry. The concept really says that similar clusters with the same polyhedral electron counts may occur for diverse elements around the periodic table that have similar orbital basis sets. Cluster charges that would so apply to all examples of classical closo, nido and arachno homoatomic cluster ions for the triel (group 13, Al–Tl), tetrel (14, Si–Pb) and pnictogen (15, P–Bi) elements are listed in Table 1 [18] for general reference.

It is experiment results that show these possibilities represent some degree of fact. To this point, Table 2 summarizes all of the considerable collection of known naked clusters that follow Wade's rules, and the compounds in which these occur. The 12 different charge types and 30 compounds span a wide range of synthetic conditions, from molecular solvents to melts and neat intermetallic systems. Examples of all but the first two arachno types in Table 1 are to be found here. On the other hand, aluminum seems devoid of isolated clusters, and phosphorus examples are more molecular and generally electron-rich.

Table 3. Cluster ions with rational relationships to classical polyhedral electron counts

Ion	Compound	Symmetry	Skeletal Electrons		Nature	Ref.
Bi_9^{5+}	Bi_6Cl_7	~C_{2v} (closo)	22	$2n + 4$	Filled a_{2u} HOMO, elongated cluster	8, 40
	$Bi(Bi_9)(HfCl_6)_3$	D_{3h}				17
Ge_9^{3-}	$(K\text{-}cp)_3(P(Ph)_3)Ge_9$, $(K\text{-}cp)_3Ge_9$	D_{3h} (closo)	21	$2n + 1$	Half-filled HOMO	41, 42
Sn_9^{3-}	$(K\text{-}cp)_3Sn_9$					43
Pb_9^{3-}	$(K\text{-}cp)_3Pb_9$					42, 37
Tl_{13}^{10-}	$Na_4A_6Tl_{13}$ (A = K, Rb, Cs)	T_h (~I_h) (closo)	25	$2n + 1$	Open HOMO	44, 45
Tl_{13}^{11-}	$Na_3K_8Tl_{13}$	D_{3d}(closo)	26	$2n + 2$	Classical Tl_{12}^{14-} +centered Tl^{3+}	45
$Tl_{12}M^{m-}$	$Na_{15}K_6Tl_{18}H$ (M = Na, m = 13)	T_h (~I_h) (closo)	26	$2n + 2$	Tl_{12}^{14-}+ centered	46
	$Na_{14}K_6Tl_{18}$ M (M = Mg, Zn, Cd, Hg, m = 12)				Na^+, Mg^{2+}, etc.	33

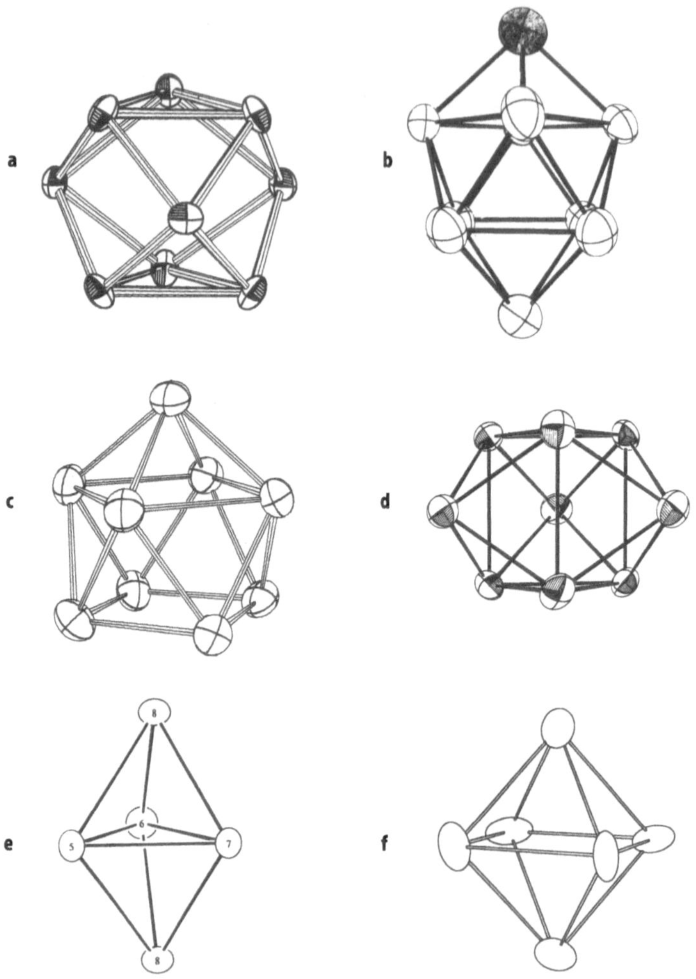

Fig. 1. Some relevant heavy element clusters: **a** Bi_9^{5+}, **b** $TlSn_9^{3-}$, **c** Sn_9^{4-}, **d** Sn_9^{3-}, **e** Tl_5^{7-}, **f** Ga_6^{8-}

Table 3 lists five additional cluster types that can be explained as rational variants of Wade's rules, including two types that contain open shell HOMOs. Several of the anions and Bi_9^{5+} are illustrated in Fig. 1. More explanations of the synthetic routes as well as descriptions and discussions of the geometries will follow. As an aside, it will be noted from Table 1 that only *nido*-Pn_4 and *arachno*-Pn_6 are neutral possibilities rather than ions, the former appearing as P_4, Sb_4, etc. while the latter seem to be unknown. This means there are severe limits on what species with this sort of special electronic stabilities can be observed in beams of neutral homoatomic clusters (see Sect. 2.5.). There are some advantages to the fact that the cluster ions have their stabilities limited by disproportionation reactions (e.g., $Bi_5^{3+} \rightarrow 4Bi + Bi^{3+}$; $3Bi_4^{2-} \rightarrow 10Bi + 2Bi^{3-}$) rather than just rearrangement processes ($Sb_4 \rightarrow 4Sb$(cryst)).

The character of two of the early examples of bismuth cations, Bi_8^{2+} and Bi_9^{5+}, bear further comment. The closo clusters would be Bi_8^{6+} and Bi_9^{7+}, respectively, but examples with such relatively large (formal) charges (fractional oxidation states) have to date been found only as anions in "neat" systems with small cations, principally for triel elements (Table 2, 3; Sect. 3). In other words, the location of the effective nonbonding energy level for a cluster compound must to some degree depend on ion charges and the coulomb energy, and polybismuth cations stable with $AlCl_4^-$, $HfCl_6^{2-}$, $BiCl_5^{2-}$, etc. anions exist only with lower charges. (The same applies to polyanions with metal–crypt cations, below.) Given an octabismuth species, the only high symmetry alternate to Bi_8^{6+} is the square antiprismatic *arachno*-Bi_8^{2+}. Second, the observation of Bi_9^{5+} (Table 3, Fig. 1a) with an apparent closo configuration but 22e ($2n+4$) skeletal electrons was for some time viewed as simply a violation of Wade's rules. Even so, the geometry always seemed noteworthy in that the trigonal prism was elongated by 15% relative to the length of the basal edge, contrary to the proportions observed for $B_9H_9^{2-}$ and, subsequently, for other species. It was later realized that the dimensional reapportionment accompanying this "violation" reflects an alternate stabilization (see 2.3.3) relative to the normal 22-electron nido prospect found for several Tt_9^{4-} as unicapped square antiprisms. An appropriate caution here; any supposed explanation for the absence of any particular ion or compound on some absolute basis without detailed knowledge of all other possibilities is probably specious; what is really illustrated is that another phase is instead more stable thermodynamically or kinetically or, perhaps, less soluble.

Attempts to extend these ideas to polyantimony cations have resulted in only "dry runs". The inherent reducibility of the molecular $SbCl_3$ by Sb is very small, and addition of Lewis acids such as $AlCl_3$ to "free" the cation from complexation has demonstrated some usual solution effects, but it is unhelpful in providing any reduced phases [11,47]. On the other hand, numerous polycations of the chalcogen elements (Ch) may also be obtained with suitable Lewis acids, $Se_8^{2+}(AlCl_4^-)_2$ [23], $Se_{10}^{2+}(AsF_6^-)_2$, and $Te_6^{2+}(AsF_6^-)_2$, for example [48]. However, these as well as their polyanions are not pertinent to our present considerations as they are all electron-rich and contain localized electron pairs with stereochemical impact. Generally, clusters with many more than 3 *p* electrons per main-group element fall in this category [38]. The arachno cations in $Ch_4^{2+}(AlCl_4^-)_2$, etc. are exceptions as these present delocalized p HOMOs. Contrary to general expectations for boranes, the Ch_4^{2+} and Bi_8^{2+} members show the loss of trans vertices from the closo parents, while it is the lower order vertex that is removed in the discriminating nido examples Tt_9^{4-}.

The majority of the compounds in Tables 2 and 3 that illustrate the generality of Wade's rules are not the "acid-stabilized" cationic members but rather those that contain polyanions. One is a group of the Zintl anions isolated with well complexed alkali-metal cations, and the other type occurs in "neat" alkali-metal–triel systems. These will be described separately.

2.2
Zintl Anions with Alkali-Metal–Crypt Cations

The expansion of these concepts from isolated novelties with bismuth to an extensive general chemistry of naked polyhedral anions that fit well with classical bonding models was first achieved through stabilization of compounds of the many so-called Zintl ions and discovery of several new members as well. This nomenclature recognized Zintl's extensive electrochemical and extractive studies of main-group polyanion solutions in liquid ammonia, mainly in the 1930s, for Pb_9^{4-}, Sn_9^{4-}, Sb_7^{3-}, Bi_3^{3-}, Bi_7^{3-}, and so forth (see [3] for the history). These solutes were obtained by cathodic reduction of the elements or via extraction of a suitable alkali-metal alloys or their mixtures. Alkali-metal compounds of the heavier group 13 (triel) elements and of Si and Ge did not yield significant solutions. It is important to note that none of the solid intermetallic precursors contains any of the clusters deduced in solution. Furthermore, none of Zintl's polyanion solutes could be isolated from liquid NH_3 and thence characterized, presumably because of inadequate stability of the $Na(NH_3)_n^+$ etc. countercations relative to the binary alloys. The utilization of crypt (or cryptand, abbreviated cp) ligands, cyclic polyether amines, to afford more stable complexes with the alkali-metal cations enabled the isolation of a large number (but not all) of Zintl's reported polyanions as well as some new ones, namely (Sn or $Pb)_5^{2-}$, $(Sb,Bi)_4^{2-}$ and $Ge_9^{2-,4-}$ [49]. (Some success had been achieved previously with ethylenediamine, but its complexes are less stable and the cations were not always well sequestered [50].) The effectiveness of this method can be viewed in two ways: 1) large counterions generally repress the disproportionation of polyatomic partner ions, as $AlCl_4^-$ does for Bi_5^{3+}; 2) sequestration of the alkali-metal cation by crypt eliminates, or greatly reduces, the delocalization of electrons from the polyanion back onto the cation that could result in a more intermetallic product, as in the disproportionation of Zintl's lead polyanions into the element and $KPb_{\sim 2.2}$, the latter falling within the homogeneity range of an inverse Cu_3Au-type phase.

Several of the unisolated members on Zintl's list have higher charges per atom than found among the crypt isolates, e.g., Pb_7^{4-}, Pn_3^{3-}, Pn_5^{3-}, and so other solvents and cations may still prove useful for their isolation. The first two would be plausible nido and arachno species, respectively. Although Pn_5^{3-} is not a classical possibility, the analytical data reported for the bismuth product are actually close to the composition of the known Bi_4^{2-} [26]. The $(A\text{-}cp^+)_3Sb_7^{3-}$ examples are not considered here as this C_{3v} anion has formal lone pairs on the three two-bonded Sb atoms (equivalent to sulfur in P_4S_3). The apparently lower charges per cluster atom that seem necessary to gain crypt-cation derivatives are also nicely illustrated by tetrahedra clusters. Recall that Tt_4^{4-} clusters are found in the neat alkali-metal–tetrel systems (Table 2), but not in the crypt systems. However, lower charged isoelectronic products with the mixed $Tt_2Pn_2^{2-}$ clusters listed in Table 2 are obtained as K-cp^+ salts, viz., $Sn_2Bi_2^{2-}$, and many others are presumably probable. The mixed "tetrahedra" in these phases are more-or-less disordered within the rather uniform cavities provided by the large cations, an unfortunate event that occurs in some other instances as well (below). Increasing dissimilarity between the component elements in a mixed cluster will eventually lead to cluster configurations that distinguish them, e.g., the isoelectronic but butterfly-shaped $Tl_2Te_2^{2-}$ [51].

A greater breadth of stability is found for five-atom polyhedra, where isoelectronic trigonal bipyramidal ($\sim D_{3h}$) clusters span Tr, Tt and Pn families. The nominal Tl_5^{7-} polyhedra (along with others) are found in fairly complex alkali-metal salts such as $Na_{23}K_9Tl_{15.3}$ (Fig. 1e), the Sn_5^{2-}, Pb_5^{2-} analogues occur as Na-cp$^+$ salts, and Bi_5^{3+} is captured under acidic conditions as the $AlCl_4^-$ salt. Other triel polyhedra found only in neat systems are often rather different, and these will be elaborated in Sect. 3.

2.3 Nine Atom Polyhedra

2.3.1 *Synthesis and Structures*

The greatest diversity and interest occur among the Tt_9 clusters, all with crypt–alkali-metal cations (Tables 2, 3). The most direct results came first from the tin systems, basically by extraction of ASn_x alloy mixtures ($\sim 0.5 < x < \sim 2.2$ and A = Na or K) into ethylenediamine (en) in the presence of, usually, 2,2,2-crypt (Fig. 1c,d). The tin-richer systems readily yield the Sn_9^{4-} salt on solvent evaporation, and the more reduced (A-richer) alloys give the Sn_5^{2-} compound on standing. The development of finely divided tin appears to facilitate the reactions, as is formed via slow oxidation of the alloy by en or even from added NaOH. Excess crypt was originally added to avoid the potential formation of supposed double salts with $A(en)_x^+$ cations, but this unknowingly obscured other products. For instance, reduction of the relative amount of the complexing agent (by accident) resulted in a salt in which one-fourth of the potassium was not bonded to crypt, but to the anions in a $^1_\infty[KSn_9^{3-}]$ chain in which the one unsequestered cation bridges between side faces of unicapped antiprismatic Sn_9^{4-} clusters. Distortion of the latter because of this new functionality is surprisingly small. A different manipulation of the extractant from solid KSn_2 gives the surprising paramagnetic Sn_9^{3-} (below). As an aside, it is noteworthy that some chromium carbonyl derivatives of these clusters, although no longer "naked", do fit into Wade's rules electronically. All have been characterized crystallographically as the K-cp$^+$ salts. The $Sn_9Cr(CO)_3^{4-}$ ion occurs with the electrophile $Cr(CO)_3$ bound on the open square face of $Sn_9{}^{4-}$ to give the corresponding 22-electron closo heterometallic, and the $Pb_9Cr(CO)_3^{4-}$ version is also known [52,53]. The nucleophilic character of the skeletal electrons on this open square face is quite analogous to that known for metallocarboranes, in $Fe(C_2B_9H_{11})_2^{2-}$ for example [54]. Also, the otherwise unknown *closo*-Sn_6^{2-} has also been captured with $CrCO_5$ bound exo at each vertex in the novel $Sn_6(Cr(CO)_5)_6^{2-}$ [55].

The chemistries of the naked clusters of the congenors lead and germanium are not very similar. A crystalline Pb_5^{2-} salt was obtained only when the en solutions were heated [27], while the probable crystals of this phase formed at room temperature do not seem to diffract [37]. A solute species attributable to Pb_5^{2-} has not been seen in the NMR, either. Remarkably, an $(A\text{-}cp^+)_4Pb_9^{4-}$ compound of the best known (most often cited) solution cluster has never been found, but the crypt-poorer $(K\text{-}cp)_3KPb_9$, isostructural with the above tin compound, has recently been obtained when the ligand was substoichiometric relative to potassium [37].

The germanium ions are somewhat more diverse and, in part, perhaps less certain. Two structures with a pair of independent Tt_9 clusters and six cation complexes have been found. A triclinic structure was analyzed with independent Ge_9^{2-} and Ge_9^{4-} ions. Although the quality of the solution was not great [(R(F) = 0.149, R_w = 0.169)], standard deviations in Ge–Ge distances were only $(4-7)\times10^{-3}$ Å. The Ge_9^{2-} ion was seemingly closer to D_{3h} (closo) (below) but with one prism edge inexplicably 0.34 Å longer than the other two. This is the only 20-electron Tt_9^{2-} ion reported so far, although $TlSn_8^{3-}$ is a close analogue (below). It is also reassuring that studies of two crystal structures in which the anion could not be resolved well have clearly shown two cations per cluster and, in one case, an A:Ge analyses near 2:9 [56]. On the other hand, the second monoclinic $(A\text{-}cp)_6Sn_{18}$ analogue has been interpreted in terms of only Sn_9^{3-} ions (below).

It should be noted that the report of a classical bicapped antiprismatic (D_{4d}) Ge_{10}^{2-} ion came from a trigonal structure determination in which the four equivalent $K\text{-}cp^+$ cations refined well, but the anion was deduced to be three-fold disordered (trilled) about the principal axis of the cell. Precession photographs strongly supported the trigonal cell choice. Errors in the resulting anion distances were larger, ~0.01 Å (R = 0.045 for data to $2\theta \leq 44°$), but the square antiprism faces deduced were quite close to planarity. The result is plausible, but the uniqueness of the model is not completely clear.

Substitution of small amounts of thallium into tin clusters, with the intent to raise the charges on the large closo clusters and thus to separate them better by more cations, led to a structure in which the distribution within a single cluster site could be readily analyzed in terms of superimposed $TlSn_8^{3-}$ and $TlSn_9^{3-}$ units [38]. These have Tl substitution at an equitorial face-capping site in a D_{3h} polyhedron in $TlSn_8^{3-}$ and on one axial (square-face-capping) position in the nominal D_{4d} (bicapped antiprismatic) $TlSn_9^{3-}$ (Fig. 1b). (The heteroatoms are of course neglected in the symmetry designations.) These two ions appeared superimposed in equal amounts such that the thallium atoms in each appeared at the respective axial positions in a mixed bicapped square antiprismatic image. Thus, seven atoms were identical in both, while the two axial atom positions refined as the fairly unambiguous 50% Tl (from $TlSn_8$) plus 50% Sn at one site and only 50% Tl (from $TlSn_9$) at the other.

Finally, the family of naked nine-atom tetrelide clusters needs to be completed with a description of the remarkable and unexpected Tt_9^{3-} radical members (Table 3). The triclinic $(K\text{-}cp)_3Sn_9 \cdot 1.5$ en was first obtained in high yield (80–95%) under a specific set of conditions, namely, after decantation of the blood-red solution formed by the action of cp in en from the KSn_2 alloy after days to weeks; otherwise, the Sn_9^{4-} compound was obtained after several weeks [57]. The Ge_9^{3-} ion was first discovered in another triclinic cell in which a $P(Ph)_3$ molecule was present as well, evidently only in a space-filling role. The germanium and lead examples $(K\text{-}cp)_3Tt_9 \cdot 0.5en$ recently reported (Table 3) are probably isostructural with the tin compound, although these were described with a different (better) triclinic cell than the tin member. (Data for the three were all collected at different temperatures as well.) It is noteworthy that the green solution first associated with Pb_9^{4-} by Zintl and others is in fact the color of Pb_9^{3-}, while Pb_9^{4-} is more brown or brown-red [37]. All of these Tt_9^{3-} radical ions have been clearly confirmed by esr solution measurements along with esr or magnetic susceptibility studies on the solids.

2.3.2
Geometric Features

The family of the Tt_9^{n-} and Bi_9^{5+} ions show a variety of geometric differences and distortions. It is important to realize that 22-electron members (Sn_9^{4-}, Pb_9^{4-}) are known to be fluxional on the nmr time scale, as first predicted [35] and then revealed by the seminal NMR experiments of Rudolph et al. [58]. The pathway for this is likely the $D_{3h}-C_{2v}-C_{4v}$ process (along normal vibration coordinates) that converts a tricapped trigonal prism into a unicapped antiprism, and vice versa [59]. The end points and the distinguishing features can be seen in Fig. 2. The higher symmetry prismatic figure (left) can be smoothly converted into a unicapped square antiprism if one of three folded quadrilateral faces defined by waist cap–prism height–waist cap (e.g., 7-3-6-9) becomes square and planar (plus other small distortions). One prism height elongates and the cap-to-cap distance shortens to become the equivalent diagonals of the square base while the third face-capping atom turns into the axial atom capping the other side of the antiprism. Another feature is that the other two trigonal prism heights (h, e.g., 2–5) and the two interconnecting basal edge lengths (e.g., 1–2) define what becomes the ideal square face that is capped. The h–e data for a variety of Tt_9^{n-} and Bi_9^{5+} are listed in Table 4, first as two heights (prism edges) of the best trigonal prism (h_2, h_3) relative to the shortest height (h_1) in each and second, as the ratio of the average prism height h to the average basal edge e. The first gives some measure of the early stages of the distortion from D_{3h} to C_{2v} while h/e pro-

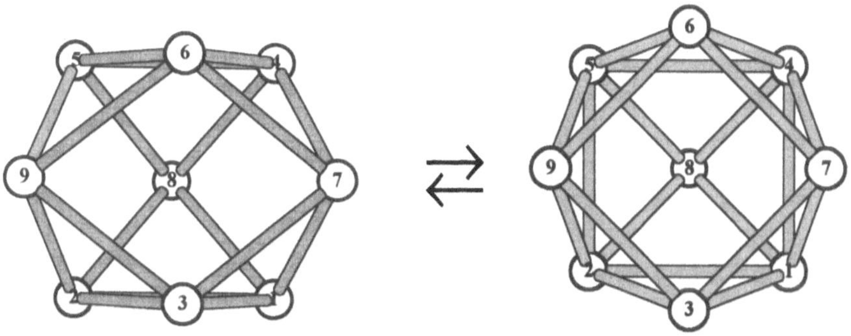

Fig. 2. A schematic of the fluxional pathway between, *left*, the tricapped trigonal prism (D_{3h}) and *right*, the unicapped Archimedean antiprism (C_{4v})

Table 4. Relative distortions in some nine-atom clusters. (a) Heights of the nominal trigonal prism relative to the shortest (h_1), (b) Ratio of the average height (h) to the average basal edge (e) of the trigonal prism.

		Bi_9^{5+}		Sn_9^{3-}		Pb_9^{3-}		$TlSn_8^{3-}$	Ge_9^{2-}	Ge_9^{3-}		Sn_9^{4-}	
(a)	h_3	1.00	(1.07)	1.02	(1.10)[a]	1.06	(1.08)[a]	1.05	1.13	1.16	1.16	1.30	1.30
	h_2	1.00	(1.00)	1.02	(1.02)	1.00	(1.02)	1.00	1.02	1.13	1.10	1.01	1.02
(b)	h/e	1.15	(1.19)	1.08	(1.08)	1.07	(1.06)	1.03 (1.01[b])	1.10 (1.06[b])	1.17	1.19		
ref		17	8	43	60	42	60	38	34	41	42	35	36

[a] The undisordered cluster in the $(K\text{-}cp)_6Tt_{18}$ phases.
[b] Neglecting one long prism edge.

portions will be useful in correlating an alternate ~D_{3h} distortion when 20 electron clusters are reduced to Tt_9^{3-}, Bi_9^{5+}, etc. Some clusters are not really on the idealized pathway, e.g., Ge_9^{3-} where two edges are elongated. Inclusion of Sn_9^{4-} (~C_{4v}) in Table 4 is only to define the relative "prism" dimensions at the other end of the transition where one height has become the base diagonal (Fig. 2).

Distinguishing well among approximate D_{3h} polyhedra, C_{2v} intermediates, and other distortion directions is accomplished more precisely with the aid of the more discriminating dihedral angles (δ) [59]; illustrations alone can be misleading. Those angles that are especially diagnostic (see Fig. 2) are those between (a) the basal faces in the trigonal prism (1-2-3, 4-5-6; 180°) relative to opposed faces in the waist of the antiprism (~160° in these examples) and (b), the vicinal cap-to-cap folds about the prism edges in the former (7-3-6, 9-3-6), one of which becomes the square base ($\delta = 0$). The other two cap-to-cap angles plus two of six vicinal prism end-to-cap angles in the tricapped prism become the four equal cap-to-waist angles in the antiprism. The course of the other four is not very discriminating.

These dihedral angles are listed in Table 5 for all of the well-defined nine-atom clusters described above. For the most part, rather smooth transitions appear to be present along the pathway imagined in Fig. 2. The Bi_9^{5+} cation, which has crystallographic C_{3h} symmetry but is insignificantly different from D_{3h} [17], provides some guidelines for normal vicinal angles, but these also depend somewhat on h/e. (The original Bi_9^{5+} in Bi_6Cl_7 [8] can be described as being about one-fourth of the way along the C_{2v} distortion path.) The radical ions Sn_9^{3-} and Pb_9^{3-} are clearly close to the D_{3h} limit, but angles in the two Ge_9^{3-} examples correspond to clusters that are on the order of one-third to one-half of the way along the transformation, but irregularly so since *two* prism edges are elongated (Table 4). The sizable but "regular" distortion of Ge_9^{2-} from the predicted closo (tricapped trigonal prism) polyhedron is also inexplicable. However, the isoelectronic $TlSn_8^{3-}$, with Tl substituted in a waist-capping position, is close to ideal. The first two angles for this in Table 5, 177° and 16°, demonstrate that Tl does not open up the capped face; in fact, these prism heights are 0.15 Å (4.3%) less than the third. The remainder of the angles in this ion are not listed because of the complications generated by the larger thallium.

More evidence of flexible or disordered clusters in the solid state have been uncovered by Fässler and Hunziker in the monoclinic structure of $(K\text{-}cp)_6Tt_9Tt_9$ compounds, Tt = Sn, Pb, which are paramagnetic and esr-active [60]. One of the two independent clusters in each example refined in a straightforward manner, and data for these in Tables 4 and 5 show that their symmetry is not far removed from D_{3h} and from the cluster characteristics in the separately defined $(K\text{-}cp)_3Tt_9$ phases for Sn and Pb. On the other hand, the Fourier map showed that the cluster at the other site exhibited bean-shaped ellipsoids for two of nine tin atoms (three for lead), the midpoints of which turn out to be adjacent positions in the base of a plausible ~C_{4v} antiprism. However, in both cases these strange ellipsoids could be deconvoluted well into 50:50 pairs so as to define two ~D_{3h} antiprisms rotated 90° from each other about a proper two-fold axis. The folded face behind (Fig. 2, left) comes out remarkably square in this process, corresponding to h/e = 1.02. The largest prism height ratios (h_3/h_1) in the disordered

Table 5. Some dihedral angles (δ, deg) in nine-atom polyhedra[a]

D_{3h}-type-faces		trigonal prism (opposed) 1-2-3, 4-5-6	cap to cap (vicinal) 7-3-6, 9-3-6						prism end to cap (vicinal) 1-2-3, 9-2-3				
D_{3h}	Bi_9^{5+}	180	22 (x3)[b]									43 (x6)[b]	
$\sim D_{3h}$	Sn_9^{3-}	179	17	18		18			40		40	4 @	41–42
	$TlSn_9^{3-}$	177	16										
	Pb_9^{3-}	176	14	20		21			38		39		43–44
	Pb_9^{3-} [c]	176	16	19		23			38		39		42–43
	Sn_9^{3-} [c]	174	13	19		23			37		38		42–44
C_{2v}	Ge_9^{2-}	171	8	23		25			38		36		44–47
	Ge_9^{3-}	170	16	18			31						
	Ge_9^{3-} [d]	165	14	17			32						
$\sim C_{4v}$	Ge_9^{4-}	162, 156	5		24		32	32		33			51–54
	KSn_9^{3-}	160, 156	2		28		29	31		31			51–54
C_{4v}	Sn_9^{4-}	158	3		29		30	29		32			52–55
	KPb_9^{4-}	159	1		29		31	31		31			52–54
C_{4v}-type faces		waist opposed	base		cap to waist, vicinal							waist to waist, vicinal	

[a] References in Tables 2 and 3 unless noted otherwise.
[b] Angles depend on h/e (Table 4).
[c] The bettter behaved cluster in $(K\text{-}cp)_6Tt_9Tt_9$ cell, ref 60.
[d] Ref 41.

pair are 1.20 and 1.14 for tin, 1.18 and 1.09 for lead, somewhat larger (more distorted) relative to both 1.10 and 1.07 for the other well-behaved cluster in each structure and 1.02 (Sn) and 1.06 (Pb) for single Tt_9^{3-} clusters in the analogous $(K\text{-cp})_3Tt_9$. The magnetic moments have been found to change appreciably with temperature, further complicating the detailed picture [61]. These results led them to the suggestion that the earlier analysis of $(K\text{-cp}^+)_6Ge_9^{2-}Ge_9^{4-}$ in a different structure type might also be interpretable in this way, although the ellipsoids in the latter were not so distinctly misshapen. (Unfortunately, there was no impetus at the earlier time to examine magnetic or esr data for the compound.) The higher charged ion assigned as Ge_9^{4-} does have more cation neighbors, however, 16 vs 13 out to 13.5 Å. It remains to be seen what other problems may appear in the sometimes rather unrestrictive cavities present in the A-cp$^+$ sublattices; several other structures have been encountered in which evident cluster disorder of apparent nine-atom clusters could not be completely resolved, including some that clearly had two or four A-cp$^+$ cations per cluster.

2.3.3
Calculations

Theoretical treatments, most at the fairly simple EH level, have not only generally supported what has been observed in closed shell configurations but have also offered a useful explanation of the Bi_9^{5+} "violation" [62,63]. But no calculations have been made at a thorough enough level to even approach accounting for absolute stabilities, rather these relate only the one-electron internal energies and electronic structures of the isolated clusters. (Some species are naturally not bound under these conditions.) Even at the extended Hückel level, the relative proportions of the clusters can be important for consistent results. Agreements reported between calculations and observed closed-shell geometries for Tt_4^{4-}, Pn_4^{2-} and Ch_4^{2+}, Tt_5^{2-} and Bi_5^{3+}, Tr_6^{8-}, and Bi_8^{2+} (Table 2) are all rather straightforward and will not be detailed. Most of the centered icosahedral thallium cluster examples in Table 3 represent new geometric configurations but nothing very unusual in bonding. The classical $2n+2$ icosahedron would be Tl_{12}^{14-}, but such a large polyhedron would presumably be poorly bonded as only centered examples are known. Addition of the centering Tl^{3+} introduces no new cluster bonding orbitals or electrons (one filled a_{1g} is already present) and produces the observed Tl_{13}^{11-}. Analogous products involving centered Na^+, Mg^{2+}, Zn^{2+}, Cd^{2+} or Hg^{2+} are also known. The specific crystal chemistry and cation packing that make the triel systems so interesting and a few observations on relativistic effects as well will be considered in Sect. 3.

The nine atom polyhedra described in Tables 4 and 5 need some attention, particularly as to fluxionality and the cluster proportions. The intracluster fluxionality of Sn_9^{4-} and Pb_9^{4-} first observed by NMR by Rudolph et al. [58] seems quite consistent with the process shown in Fig. 2. Ions with ~C_{4v} configurations evidently lie somewhat lower in energy in the solid state (or, at least potentially, form less soluble salts). The energetic plausibility of the pathway has been considered for Sn_9^{4-} by Burns and coworkers [63] in terms of symmetry and normal vibrations of the end members. The low frequency vibrations of Pb_9^{4-} and its softness to deformation have been studied by density functional theory as well [37]. In contradistinction, a comparable fluxionality of 20- or 21-electron D_{3h} clusters is not likely because the resulting HOMO's for the C_{4v} clusters would be e^2 or e^3 [59]. The former Tt_9^{2-} species have not been observed for certain in NMR solution studies, and the latter Tt_9^{3-} are of course paramagnetic and would give only very broad signals, if any.

A range of ~16% in h/e among known nine-atom polyhedra listed in Table 4 that are approximately D_{3h} in symmetry seems to be especially significant, and lack of some care and caution regarding known or hypothetical species can greatly effect the credibility of MO calculations. The origin of this variation in relative prism height arises with an a_2 orbital which becomes the HOMO on transition from 20- to 22-e clusters. This MO primarily of p_z (axial) orbital contributions that are σ antibonding along the parallel edges (heights) of the trigonal prism but π bonding within the basal faces; hence an increase in h/e on reduction is expected as h and e vary in opposite directions [64, 65, 57]. This is in fact manifested by a fairly linear increase in h/e with increasing electron count from $TlSn_9^{3-}$ (and also $B_9H_9^{2-}$) through Sn_9^{3-} and Pb_9^{3-} to Bi_9^{5+} (Table 4) (3, 43). Ge_9^{2-} and Ge_9^{3-} are out of line in this respect. This, of course, does not provide a sufficient reason why Bi_9^{5+} does not have a conventional C_{4v} configuration instead. However, it has been noted that the tricapped trigonal prism also affords the best minimization of internal charge repulsions between the cluster atoms, i.e., the points-on-a-sphere distribution, and this is even better when the prism is elongated [63]. It is important to note that any complete explanation must also address the effect of the dipole of the C_{4v} configuration, although more recent calculated charge distributions suggest its magnitude may be less than once thought [37,63]. Another challenge is an energetic explanation as to why the unusual Tt_9^{3-} radical ions should be stable at all relative to Tt_9^{2-} (if any) and Tt_9^{4-}.

Calculations on D_{3h} polyhedra as a function of proportions have evidently only been examined once, and that only by extended Hückel means [57]. The charge-consistent results as a function of h/e but otherwise with averages of the observed distances appear to be consistent internally inasmuch as the one-electron energy sums results agree with observations, although by very small margins in some cases. These are, with the energy for the observed proportion in bold type, Bi_9^{5+}: −737.9, −739.2, **−740.1** eV for h/e = 1.01, 1.08, 1.15, respectively; Sn_9^{3-}: −338.2, **−338.5**, −338.4; Sn_9^{2+} (hypothetical, vice $TlSn_8^{3-}$) **−365.7**, −365.5, −365.0. In each case, the HOMO–LUMO gap is also largest for the observed h/e. Similarly, Sn_9^{4-} was calculated to be 1.0 eV more stable as the observed capped square antiprism than with any D_{3h} proportion (h/e = 1.15 was the closest).

EHMO energies for Sn_9^{3-} and Sn_9^{4-} at the ideal limits for both D_{3h} and C_{4v} configurations as well as those for the unknown Sn_9^{2-} with proportions based on $TlSn_8^{3-}$ are plotted in Fig. 3 [57]. The h/e proportions and total energies were as just quoted or, for C_{4v} Sn_9^{3-}, −336.9 eV (one cycle), D_{3h} Sn_9^{4-}, −310.8 eV; C_{4v} Bi_9^{5+}, −740.0 eV. The MO energy distribution found for the hypothesized Sn_9^{2-}, Fig. 3, is unusual in that only a 1.6 eV gap separates the rather isolated a_2'' LUMO from the e' HOMO. This may be an indication of the low stability of this unknown ion or, effectively, of an increased stability for Sn_9^{3-}. Relativistic effects may be appreciable in some of these comparisons.

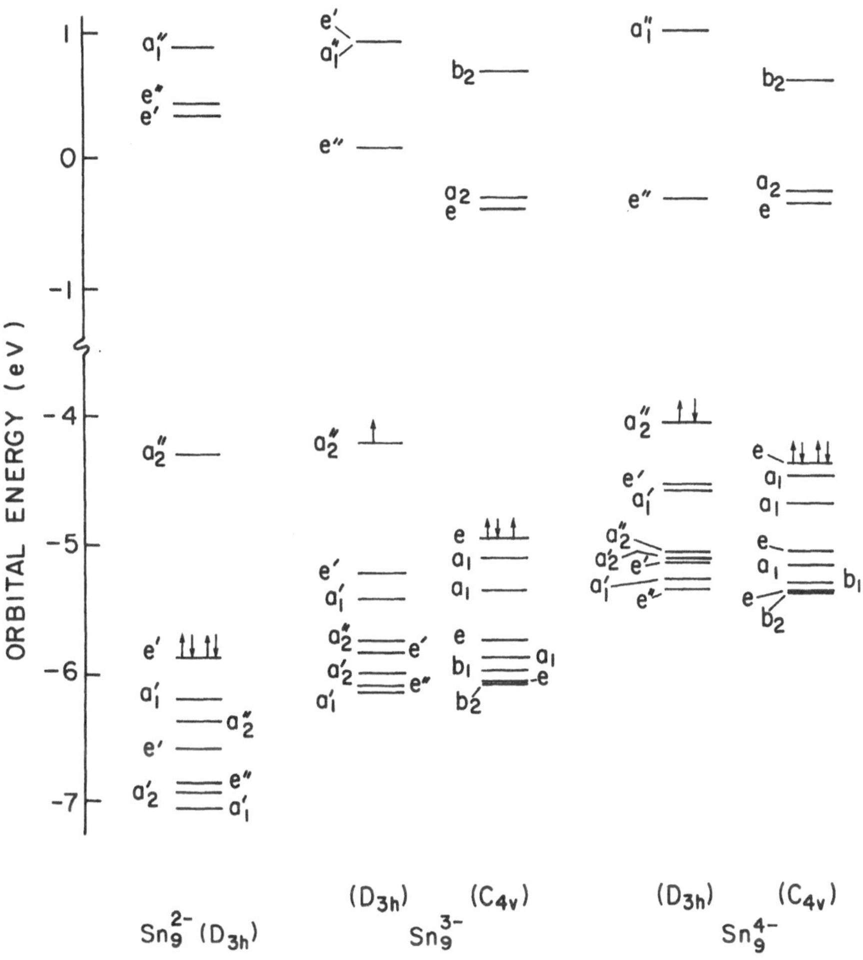

Fig. 3. EHMO orbital energies for nine-atom clusters with observed or optimized proportions. *Left*: Sn_9^{2-}, D_{3h}; center: Sn_9^{3-} with D_{3h} and C_{4v} symmetry; *right*: Sn_9^{4-} as D_{3h} and C_{4v} polyhedra. *Arrows* mark the respective HOMO in each case [57]

2.4
Other Solution Prospects

More examples or relatives of these Zintl anions are certain to be found. Even with conventional synthetic procedures, the details of not only time, temperature, solvent, and alloy but also of whether the alloy is separated from solution at some stage or not are all important. Rudolph noted early on that cluster reactions and exchange took place relatively rapidly on the alloy surfaces, but that the homogeneous exchange/equilibration reactions were quite slow [66]. A number of otherwise uncharacterized species (relevant to present considerations) have been proposed from NMR studies, for example, $TlSn_8^{5-}$ (nido?), Sn_4^{2-}

(diamagnetic, fluxional), ternary Tl–Sn–Pb clusters, Sn_8Sb^{3-} [58, 66, 67], while other ternary solutions (e.g., in K–Ga–Sn, K–In–Ge systems) have not yielded usable crystals. Changes in the cation components have also given different results, e.g., benzocrypt, Li^+ with 2,1,1 crypt, etc., presumably by affecting relative solubilities (lattice stabilities) [57]. In addition, quite a few solid products have not succumbed to structure analysis because of evident disorder or mixed species at lattice sites or, perhaps, subtle errors in the crystallography.

2.5 Metal Clusters in the Gas Phase

Recent investigations of clusters in the gas phase, usually via molecular beams, have frequently revealed the presence of a large distribution of species in any one system. These spectra also usually show a remarkable absence of marked size discrimination among the clusters of many metals, Al, Ga, Sn, Pb or Bi for instance, when either neutral or singly-charged species are examined [68, 69]. This is consistent with the generally larger negative charges that would be necessary to achieve nearly all of the nine types of closed-shell "Wade's Rule" clusters listed in Table 1. Only the classic Pn_4, Pn_3^+, Pn_5^+, Pn_7^+ would be possible for neutral or singly-charged species on the basis of these widely applicable criteria of stability. In other words, the absence of any strikingly stable cluster ions in many homoatomic beams may suggest that the many species observed are markedly less rather than only slightly less stable than ions described in this article.

The small window available when homoatomic beams are examined is nicely enlarged when mixed metal beams are employed, and strikingly different stabilities among possible species are then seen. Some of the species found (and their presumed homoatomic analogues) are $Cs_3Pb_5^+$, Pb_3Sb_2 and $Pb_2Sb_3^+$ (Pb_5^{2-}) [68–70], $Pb_4Sb_5^+$ and $Cs_5Pb_9^+$ (Pb_9^{4-}) [68, 70], $Bi_7In_2^+$ (Bi_9^{5+}) [71], $CsSnSb_3$ (Sb_4) and Cs_2Bi_4 (Bi_4^{2-}) [72]. However, an assumption that the favored shapes for homoatomic cases will carry over for all the mixed-metal cases is probably not warranted, e.g., for $Bi_7In_2^+$ (g) as an extreme case, as is already known for $Tl_2Te_2^{2-}$ (C_{2v}) [51] vs $Pb_2Sb_2^{2-}$ (effectively T_d) [20]. Most of these unipositive mixed-metal clusters would probably not be stable in the solid state either because of an insufficient number of cations to provide adequate interanion separations; at least no polyions with charges less than ±2 have been realized in solids even with the larger complex cations. The gas phase spectra suggest that some novel examples of mixed clusters might be isolated under favorable circumstances, i.e., analogues of the presumed *closo*-$Sn_3Bi_3^+$ and -$Sn_4Bi_3^+$ ($2n+2$) and perhaps of the unexpected and seemingly hypoelectronic ($2n$) species Pb_5Sb^+ and Pb_3Sb^+.

3 Clusters of the Triel Elements in Solid State Systems

The group 3 [13] triel elements afford excellent tests of the boundaries of classical deltahedral bonding, yet knowledge of all but one example is of relatively recent origin. One rationalization for the earlier lack of investigation might be that these electron-poorer elements would in many cases appear to require relatively, even improbably, high cluster charges compared with what is known elsewhere, as evident in Table 1. This could have been coupled with doubt whether the successive electron affinities might, or would, become too low to achieve such high charged ions. On the other hand, and with perfect hindsight, a good appreciation (by the right people) of the significant coulombic stabilization of cluster anions afforded by relatively small cations in "neat" binary (or higher) compounds seems to have been lacking. This last characteristic couples nicely with the fact that the higher-charged triel clusters are very good reducing agents and probably would not stable in molecular solvents such as ethylenediamine or ammonia. Significantly, until the time of the discovery of In_{11}^{7-} [73], no one appeared to have tested another important question experimentally or theoretically, whether the strongly implied minimum "wall" of $2n+2$ skeletal electrons was inviolate.

With the same hindsight, there were hints of things to come. The structure of Na_2Tl published in 1967 [22] revealed isolated nominal Tl_4^{8-} tetrahedra in a diamagnetic phase. This study followed on an empirical MO treatment of the isolated square planar mercury units in Na_3Hg_2, speculating that this could be a Zintl phase with a closed shell anion Hg_4^{6-}, i.e., Bi_4^{2-} without the weakly antibonding pi (e_g^4) and less bonding e_u^4 filled [74]. That homoatomic bonding between these elements could be significant in extended solids was evidenced very early by Zintl's classic study of the NaTl (NaIn, etc.) structure, a sodium-stuffed diamond lattice $^3_\infty[Tl^-]$ [75, 76]. By the early 1980s there were also examples of alkali-metal-gallium phases containing interbonded icosahedra and other polyhedra of gallium [77, 78]. Nonetheless, the studies that led to the broad cluster discoveries described below were prompted by nothing more profound than a desire to characterize undefined binary alkali-metal–indium compounds before some ternary targets were investigated.

3.1 New Polyhedra

3.1.1 *Hypoelectronic Examples*

A surprising diversity of new chemistry and principles is found in the many cluster compounds formed between an alkali metal and, especially, indium and thallium [79]. First are the examples of classic Wade's rule polyhedra already listed and referenced in Tables 2 and 3, namely the *nido*-In_4^{8-}, Tl_4^{8-} (T_d), and In_5^{9-} (C_{4v}) together with *closo*-Tl_5^{7-} (D_{3h}), Ga_6^{8-} and Tl_6^{8-} (O_h), Tl_{13}^{11-} ($\sim I_h$) (the Tl-centered Tl_{12}^{14-}) and some $Tl_{12}M^{-14+m}$ derivatives in which m is the conventional oxidation state of M. More important are the ground-breaking examples of new cluster types listed in Table 6, namely, the first hypoelectronic ($2n-4$) clusters Tr_{11}^{7-}, the novel $Tl_9Au_2^{9-}$ derived by substitution in and distortion of the Tl_{11} cluster, the centered $Tr_{10}Zn^{8-}$ and $Tr_{10}Z^{10-}$ (Z = Ni, Pd, Pt), two examples of a Jahn-Teller-type distortion in Tl_6^{6-}, and the unusual Tl_9^{9-}. Six of these are illustrated in Fig. 4. The very important cation "solvation" and stabilization of many of the foregoing will be considered shortly. Before going further, a caution is

Table 6. Hypoelectronic, centered, and other different clusters of the Triel elements Ga, In, Tl

Ion	Compound	Ion Symmetry	Skeletal Electrons		Ref.
Hypoelectronic					
Ga_{11}^{7-}	Cs_8Ga_{11}, $Cs_8Ga_{11}Cl$				80
In_{11}^{7-}	A_8In_{11}, A = K,Rb	~D_{3h} (closo)	18[a]	$2n-4$	73, 81
$In_{10}Hg^{8-}$ [b, c]	$K_8In_{10}Hg$				82
Tl_{11}^{7-}	A_8Tl_{11}, A = K,Rb,Cs				83, 84
$Tl_9Au_2^{9-}$ [b]	$K_{18}Tl_{20}Au_3$	D_{3h}	16[a]		85
Tl_6^{6-}	KTl	~D_{2h} (compress-	12	$2n$	86
	CsTl	~D_{2h} ed O_h)			87
Tl_9^{9-}	$Na_2K_{21}Tl_{19}$	C_{2v}	18	$2n$	28
Centered[d]					
$In_{10}Zn^{8-}$	$K_8In_{10}Zn$	D_{4d}	18	$2n-2$	88
$Tl_{10}Zn^{8-}$	$K_8Tl_{10}Zn$				89
$Ga_{10}Ni^{10-}$	$Na_{10}Ga_{10}Ni$	~C_{3v}	18	$2n-2$	90
$In_{10}M^{10-}$	$K_{10}In_{10}$ M, M=Ni,Pd,Pt				91

[a] An extra electron is delocalized (see text).
[b] Substitutional heteroelement.
[c] Hg disordered over trigonal prismatic sites.
[d] Table 3 lists other centered Tl_{12} examples that are more classical.

appropriate regarding the relatively large apparent charges assigned to these cluster ions. These numbers are intended to reflect the cumulative count of valence electrons or aggregate oxidation states, and they should not be taken as real charges on the polyanions. Some polarization by, covalency with, or electron delocalization back onto the cations is to be expected. Such circumstances do not fundamentally alter the process of counting electrons in cluster orbitals in compounds with such substantial differences in valence energies as do the alkali metals and the triel elements.

The new cluster polyhedron Tr_{11}^{7-} now known for Ga, In and Tl demonstrates how this unprecedented cluster can be hypoelectronic relative to standard descriptions of deltahedra. The quite simple basis for this for the first-described In_{11}^{7-} is illustrated by the numbered polyhedron in Fig. 4a and the MO's in Fig. 5. A plausible precursor is the regular tricapped trigonal prismatic In_9^{11-}. Reduction of its charge followed by a polyhedral expansion can be imagined starting with capping the two basal prismatic faces by In^+ (In2) ions to give In_{11}^{9-} ($2n-2$). Introduction of these along the three-fold axis creates no new bonding skeletal MO's and adds no p electrons, only a pair of $6s^2$ core-like levels. The resulting polyhedron with its MO levels shown at the left in Fig. 5 is multiply convex, not very spherical, and suffers from low order bonding at the added atoms. But the observed axial compression along the three-fold axis together

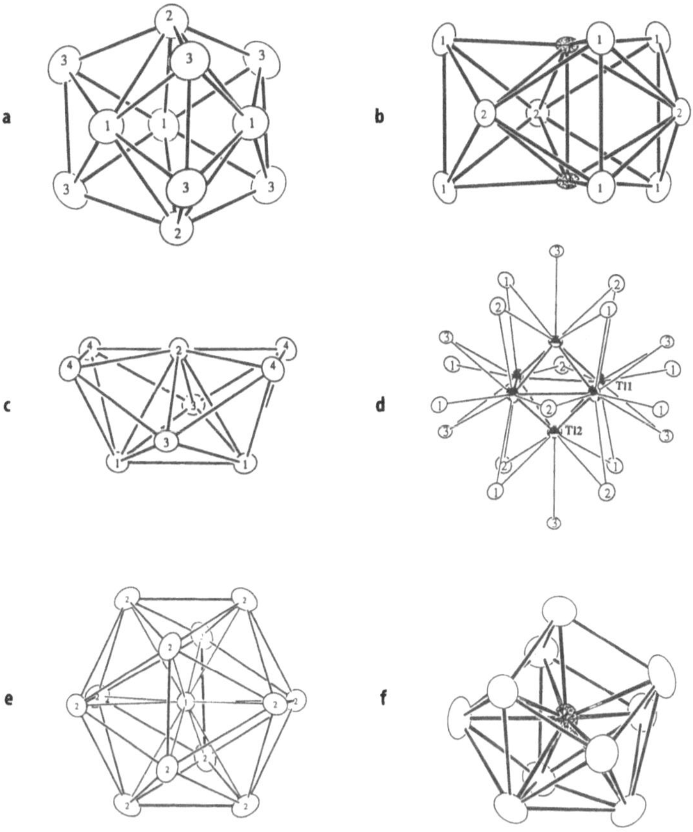

Fig. 4. Some unusual triel clusters: **a** $In_{11}7-$, **b** $Tl_9Au_2^{9-}$, **c** Tl_9^{9-}, **d** Tl_6^{6-} with Cs^+ neighbors, **e** Tl_{13}^{11-}, **f** $In_{10}Ni^{10-}$. The compounds containing these are referenced in Tables 2, 3 and 6

with radial expansion opens up the basal faces of the trigonal prism (In3) to 5.0 Å, empties the a_1' that was formerly bonding in those faces, and produces three more neighbors for each In2, the waist-capping In1. Notice also the general lowering of the energy. The observed D_3 cluster deviates from D_{3h} by only a 3.1° twist in the trigonal prism, and the In–In distances range between 2.96 Å (In1–In3) and 3.28 Å (In1–In2). This particular pentacapped trigonal prism may still be considered to be somewhat different from regular deltahedra by virtue of three surface creases defined by pairs of triangles that share In1–In2 edges (their normals intersect outside of the figure with $\delta = 21.9°$). The novel presence of one more alkali metal in the compound than necessary for the anion charge, viz., as $(A^+)_8Tr_{11}^{7-}e^-$, will be considered later (Sect. 3.2)

The unusual $Tl_9Au_2^{9-}$ ion (Fig. 4b) can be derived directly from Tl_{11}^{9-} starting with gold substitution on the two axial (Tl2) positions. The four-electron deficiency therefrom (above the low lying Au $5d^{10}$) is accommodated first by the formation of a strong transannular Au–Au bond (d = 2.96 Å), with little distor-

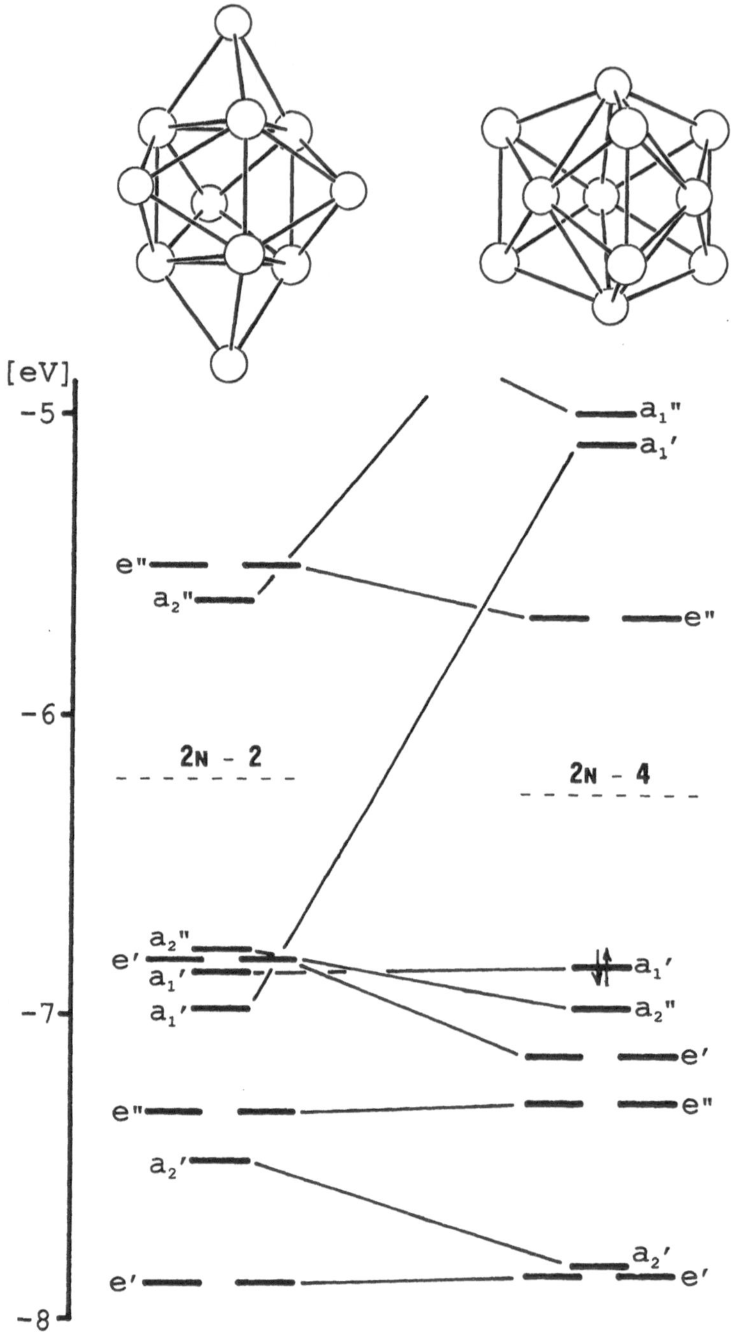

Fig. 5. MO energies for the imagined generation of In_{11}^{7-}. *left*: the result of axial capping of the tricapped trigonal prism In_9^{11-} by two In^+; *right*: after distortion to the observed geometry. Note the loss of the a_1' bonding orbital concentrated in the basal faces of the prism [73]

tion of the remaining Tl_9 portion, and second, by the addition of two electrons to the anion. In the first step, the $2a_2''$ penultimate and the $2a_1$ HOMO orbitals in the parent, both of which have substantial p_z components, are respectively lowered in energy by Au–Au bond formation and pushed up as the σ^* $2a_2''$ (out-of-phase) level. The gold atoms are effectively reduced with two $6s^2$ levels present in the lower *s*-band.

Two other cluster types that might be called hypoelectronic, Tl_6^{6-} and Tl_9^{9-}, are not quite as unusual in origin. The phases KTl and CsTl both contain Tl_6^{6-} ions, tetragonally compressed octahedra ($\Delta = 0.364$ Å in CsTl). This process splits the penultimate t_{1u} of Tl_6^{8-} into the modestly bonding e_u^4 and an empty a_{2u} (predominantly p_z), a typical Jahn-Teller distortion. The cesium salt has somewhat more regular bridging by the cations and is close to the ideal D_{4h} symmetry (Fig. 4d). These structures are interesting alternatives to the 3D structure of NaTl and appear to originate because of cation size [87]. The contrasting nonathallium cluster, Fig. 4c, is at first difficult to relate to other polyhedra. An alternate distortion of a tricapped trigonal prism, lengthening two edges to give a C_{2v} intermediate, has been suggested as a second (and evidently unrealized) fluxional process for this polyhedron [28]. In Tl_9^{9-} these edges would correspond to the two longer Tl4–Tl4 separations (5.34 Å), very large relative to Tl1–Tl1 and a considerably greater distortion than first proposed. A much more direct process comes instead through removal of the four top vertices in a Tl_{13}^{11-} cluster, Fig. 4e. (The Tl_9^{9-} cluster can thus be alternately described in terms of two fused pentagonal bipyramids defined by axial Tl1 and Tl2 and shared Tl3 (×2), Fig. 4c.) The vertex loss process leads to a much cleaner electronic description. Wade's rules don't work well when multiple adjacent vertices are removed as the skeletal pairs are not always left behind with each vertex loss [92]. Direct calculations for the icosahedral route show that three skeletal electron pairs are lost along with the four vertices (as Tl^+), viz., $Tl_{12}^{14-} \rightarrow Tl_8^{12-} + 4Tl^+ + 6e$, and this is then "centered" by Tl^{3+} (T12) to give Tl_9^{9-}. The result, with what is probably best described as nine skeletal atoms, contains $2n = 18$ bonding (p) electrons. Thus we see that the loss of multiple vertices plus some skeletal electrons is yet another way whereby high charges may be decreased and clusters stabilized.

3.1.2
Centered Polyhedra

Centering of classical "naked" deltahedra by certain heteroatoms also affords another and evidently new way to stabilize hypoelectronic clusters. The same type of process provides the means for stabilization of a large family of reduced cluster halides of the early transition metals [93]. Electronic description of the process to form the prototypical $In_{10}Zn^{8-}$, via charge-consistent EHMO calculations, starts with the classic bicapped Archimedean antiprismatic In_{10}^{12-} and centers this by Zn. The geometric and accompanying electronic changes are depicted in Fig. 6 [88]. The equiproportioned In_{10}^{12-} shell with the same average distance as observed in the product (left) requires an axial compression and a radial expansion to place all vertices substantially equidistant from the cluster

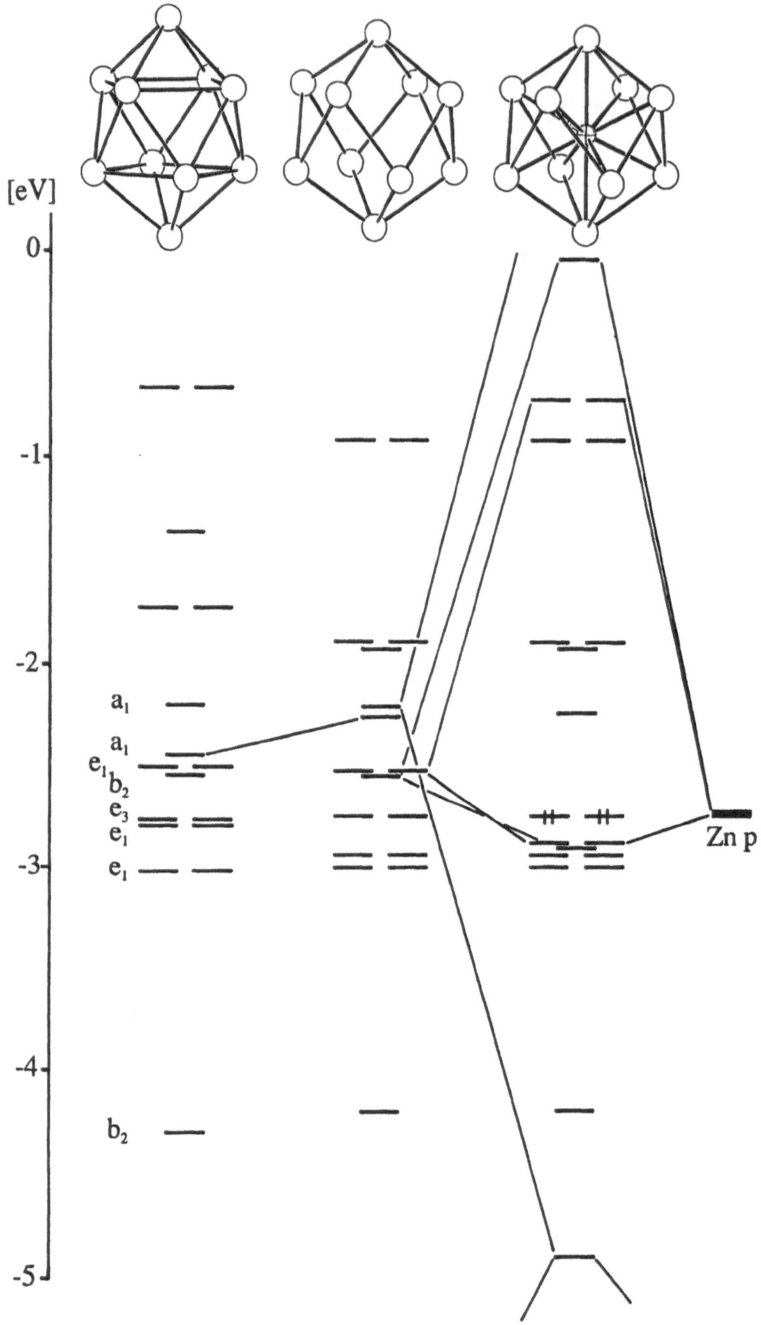

Fig. 6. MO energies changes for the formation of $In_{10}Zn^{8-}$: *Left*: the classical D_{4d} polyhedron; *center*: distortion to the observed skeletal dimensions; *right*: after zinc centering [88]

center. This change primarily opens up the capped square faces and raises the corresponding $1a_1$ level, which is both σ and π bonding therein, to a future LUMO role. The hypothetical intermediate is In_{10}^{10-} at the same gap (center). Insertion of neutral zinc ($4s^2$) causes somewhat more complex changes in the skeletal MO's than heretofore, and also has significant effects on the more core-like $5s^2$ states (right). The interstitial's s, p_xp_y and p_z ao's mix with a_1, e_1 and b_2 in the empty cluster, respectively, to stabilize the hypothetical intermediate. Three levels are obtained for each of the three representations, one of each being high lying and empty. A low-lying a_1 (–9.05 eV) contains Zn *s* and predominately In *s*, while the bonding a_1 shown near –5.0 eV is largely Zn *s* mixed with some In *s* and *p*. A low lying e_1 (not shown) is mainly In *s* (and a little In *p*) but negligible Zn *p*, while the two pair shown are, as expected, In *p* plus a greater proportion of Zn p_xp_y. (H_{ii} for In *p* and Zn *s* are about –2.40 and –6.14 eV, respectively.) A low lying b_2 at –7.70 eV has some admixed Zn p_z while the higher b_2 pair contain Zn p_z and some In *s* mixed with In *p*.

We describe the 18-e^- result as hypoelectronic relative of a more conventional 22 skeletal electron species ($n=10$), the deficiency arising equally from the distortion that empties $1a_1$ in the homoatomic parent and the two electrons contributed by zinc. Others might imagine the distorted intermediate in the center of the Figure to have the same charge as at the start (In_{10}^{12-}, without much of a gap) and to center this with Zn^{2+}, which empties one a_1 and introduces a low lying a_1 from Zn *s*. In either case, four levels are "pushed up" by the zinc, and a once-filled a_1 becomes the LUMO in $In_{10}Zn^{8-}$ because, in effect, of the cluster distortion. Incidentally, all b_2 levels shown near –4.3 eV are at the top of blocks of mainly *s*-based indium MO's and are the only ones with appreciable proportions of In *p*.

The isoelectronic centered $Tr_{10}M^{10-}$ ions for Tr = Ga, In, M = Ni and, in part, Pd and Pt offer an interesting contrast as these are geometrically closest to tetracapped trigonal prisms, the fourth cap lying on the 3-fold axis. The charge-consistent MO result (not shown) for $In_{10}Ni^{10-}$ (Fig. 4f) is similar in result to that just seen [91]. The face-opening by the capping atom and an axial compression and lateral expansion to accommodate the centering atom have comparable effects, and so does the result of the introduction of Ni in C_{3v} symmetry. Some cluster orbitals are stabilized, Ni–In bonding is introduced, and Ni *s* interacts strongly with the lowest lying In *s* (a_1) state, as before. The Ni *p* states also exhibit much the same behavior as did those for Zn, and the result is the same, a cluster with $2n$ skeletal electrons. The Ni $3d$ states appear as a narrow band near –7.4 eV, neither overlapping nor mixing with indium cage orbitals significantly. This is sensible in other terms too; Ni $3d^{10}$ is filled and low lying in such a strongly reducing cluster (~In^{-1} oxidation state).

The unusual geometric variances between these two isoelectronic clusters, D_{4d} with Zn but ~C_{3v} with Ni, have been rationalized in terms of the greater number of cations (10 rather than 8) that are bonded to the latter cluster (below). Indeed, the number of cluster edges and the average In–In distance are both larger for the ~C_{3v} nickel-centered unit, while the number of bonded indium neighbors per vertex and the total In–In bond populations are correspondingly greater for the nickel member.

3.2 Cation Effects

The roles that the cations play in the formation of the triel cluster compounds appear to specific and multiple. The qualitative effect of the coulomb potential of the alkali-metal cations in stabilizing the higher charged cluster anions seems pretty obvious relative to what is achieved with large cryptated ions and molecular solvents (Tables 2, 3, 6). It will also be noted that the clusters with the highest charge per atom, Tr_4^{8-}, have only been found with sodium, where the smaller radii of curvature for these clusters also allow for the relatively larger number of cation neighbors. The sheathing or "solvation" of triel clusters by alkali-metal cations appears to be tight and rather regular in these compounds. The cations inevitably cap triangular faces or bridge edges as well as bond exo at the vertices of the deltahedral clusters. The cations usually have a bifunctional character in bridging between, and separating, the individual clusters. (One can imagine that the clusters would condense were they not so insulated from one another.) A specificity in the cation sheathing is also supported by the observations that the cation environments about Tl_4^{8-} are very similar in both Na_2Tl and $Na_{23}K_9Tl_{15.3}$, while icosahedral clusters in several different phases are likewise surrounded by cations with very similar motifs.

Not only does solvation by the cations play a distinctive role, but so does their space filling, particularly as can be improved by mixing alkali-metal cations, the only ones (without lithium) that have been studied at all systematically. "Tuning" of structural stability with mixed cations [28] has proven to be a valuable tool in the pursuit of new phases and new clusters. The top part of Table 7 lists all of the new triel compounds that have been obtained with mixed cation compositions and which afford a good fraction of the new cluster types,

Table 7. Cation effects in the stabilization of new phases and new Triel clusters[a]

Mixed Cations Formula	Clusters
$Na_2K_{21}Tl_{19}$	Tl_5^{7-}, Tl_9^{9-}
$Na_3K_8Tl_{13}$	Tl_{13}^{11-}
$Na_4A_6Tl_{13}$ (A = K–Cs)	Tl_{13}^{10-}
$Na_{14}K_6Tl_{18}$ M, M = Mg,Zn,Cd,Hg	Tl_6^{8-}, $Tl_{12}M^{12-}$
$Na_{14}K_6Tl_{18}Na$	Tl_6H^{7-}, $Tl_{12}Na^{13-}$
$Na_{23}K_9Tl_{15.3}$	Tl_4^{8-}, Tl_5^{7-} (Tl_3^{7-}, Tl^{5-})
$Na_3A_{26}In_{48}$, A = K–Cs	Network of linked *closo*- and *arachno*-In_{12}[b]
Unmatched Cation/Cluster Charges	
1. Cation-(electron-)rich, metallic	$(A^+)_8(Tr_{11}^{7-})e^-$, A = K–Cs
	$(K^+)_{18}(Tl_{11}^{7-})(Tl_9Au_2^{9-})(Au^-)e^-$
2. Cation-(electron-)poor, paramagnetic (Curie-Weiss) clusters	$(Na^+)_4(A^+)_6Tl_{13}^{10-}$, A = K–Cs

[a] References in Tables 2, 3 and 6.
[b] Ref. 94.

seven charge types all told. In contradistinction, only the cluster types Tr_4^{8-}, Tl_6^{6-} and Tr_{11}^{7-} are found in the simpler binary alkali-metal–triel systems!

The specificity of cation interactions with the clusters and the need for good space filling are the most plausible explanations for two significant deviations of compound stoichiometry from those determined by closed-shell anion charge(s). These are listed in the bottom part of Table 7. First, the necessity for an extra cation to gain an energetically favorable packing seems to be the only rational reason for the formation of the six $(A^+)_8Tr_{11}^{7-}e^-$ examples in which an extra electron is evidently delocalized, perhaps over the cation double layers particularly. Tuning of the cluster electron count to match that of the cations in $K_8In_{10}Hg$ (in which Hg is disordered over the trigonal prismatic sites) produces only a small change in the rhombohedral space group but, more meaningfully, a switch from a Pauli-like paramagnetism for K_8In_{11} to diamagnetism in $K_8In_{10}Hg$. The gold derivative listed is also, by chance, another example of an excess of one cation and one electron per formula unit. But a possible need, somehow, for an extra electron with $Tr_{11}{}^{7-}$ ions has now been denied by the insertion of halide as well in $Cs_8Ga_{11}X$ for X = Cl, Br, I, which are now diamagnetic [80].

Two band calculations have appeared for the metallic K_8In_{11}. A charge-consistent extended Hückel study confirmed the earlier supposition that one electron per formula unit was delocalized over a VB–CB gap of about 1 eV. The DOS matched the earlier MO description of a single cluster well. The conduction band was concluded to be strongly K–In bonding, but In–In antibonding [81]. However, this conclusion may not be too firm since uncontracted atomic 4*s* orbital functions were employed for the cations, probably much too large for a condensed salt of this type. An SCF study by the periodic Hartree-Fock method and effective core pseudopotentials has also appeared [95]. The article emphasized the electrostatic origin of the cohesive energy of the solid and concluded that delocalization of the extra electron per unit over the potassium layers was consistent with the results, but not proven. The field of the cations also had a substantial and logical effect in producing a more uniform charge density within the cluster. The electron localization function (ELF) result was also analyzed. It is interesting to speculate that the localization attractors found ~1 Å outward from each indium in the cluster may represent the lone pairs.

The opposite effect, also evidently originating with tight and specific packing, is the occurrence of a shortage of cations and therefore an electron deficiency in clusters. This situation is presently known only for $Na_4A_6Tl_{13}$, A = K, Rb, Cs (Table 7). Here, a one-electron hole is localized within each $Tl_{13}{}^{10-}$ cluster, giving the compounds classical Curie-Weiss magnetic characteristics (1.74(2), 1.90(2) μ_B for A = Rb, Cs, respectively). The particularly tight sheathing of the Tl_{13}^{10-} cluster by precisely 8/2 Na atoms appears responsible, the 24 K atoms capping the other 12 faces and bonding exo at the 12 vertices. As shown in Fig. 7, a cube of sodium atoms lies on the four three-fold axes of the T_h cluster point group (and $Im\bar{3}$ space group), each Na atom thereby playing a face-capping role on two clusters. This is to be contrasted with the less versatile packing, and smaller phase field, observed for the only valence-precise analogue with a single cluster type, $Na_3K_8Tl_{11}$. Both compound types appear to be line phases without

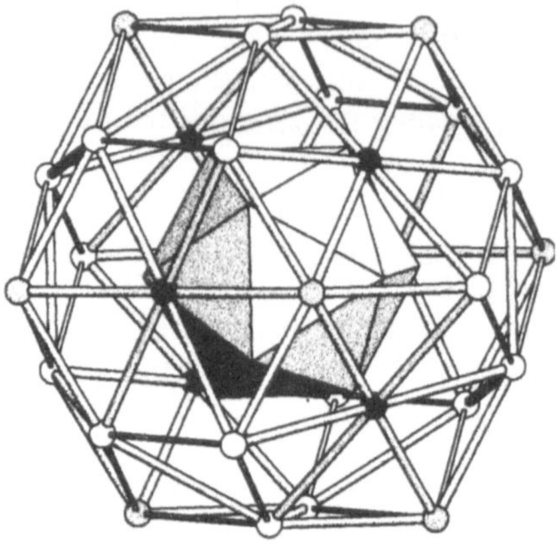

Fig. 7. A view of the In_{13}^{10-} polyhedron (*solid*) in $Na_4K_6In_{13}$ with the associated 8 Na (*black spheres*) and 24 K neighbors

significant variabilities in the cation proportions [45]. The strong and effective cation packing in $Na_4A_6Tl_{13}$ is retained in a novel family of derivatives. Simple reduction of the symmetry of bcc $Na_4A_6Tl_{16}$ from $Im\bar{3}$ to the primitive $Pm\bar{3}$ fits a group of new cluster compounds that contain equal numbers of Tl_6^{8-} (or Tl_6H^{7-}) and related $Tl_{12}M^{-14-m}$ units [33, 46]. In other words, the space group change allows one to distribute two different cluster types with the same point symmetry and total charge on the sites formerly all occupied by Tl_{13}^{10-}, and with the disappearance of the electron hole in the last as well. This change is accompanied by a switch of half of the K^+ sites into Na^+ in order to accommodate, and better solvate, the smaller octahedral clusters, which now have not only the eight face-capping sodium but also six pairs of sodium atoms at the vertices. In terms of unit cell contents this process take one from

$$Im\bar{3}:\ (Na^+)_8\,(K^+)_{12}(Tl_{13}^{10-})\,(Tl_{13}^{10-})\ \text{to}$$
$$Pm\bar{3}:\ (Na^+)_{14}(K^+)_6(Tl_{12}Na^{13-})(Tl_6H^{7-}),\ (Na^+)_{14}(K^+)_6(Tl_{12}M^{12-})\,(Tl_6^{8-}),$$
$$M = Mg, Zn, Cd, Hg,\ \text{and even}$$
$$(Zn)_{14}(Mg)_6(Zn_{13})\,(Zn_6).$$

The last step to the isostructural Mg_2Zn_{11} (plus $Mg_2Al_5Cu_6$ and $Na_2In_5Au_6$ examples as well) takes us well into the mysteries of intermetallic compounds in which the clusters are no longer clearly isolated by electropositive heteroatoms.

As might be expected, interstitialization of the thallium icosahedra to form Tl_{12} M species again produces substantial changes in two radial a_g orbitals and in the frontier t_u representation that follow the s and p orbitals, respectively, on the centered atoms [45]. Application of a relativistically parameterized EHMO treatment (REX) [96] yielded basically the same bonding picture for Tl_{13}^{10-} as without. The spin-orbit coupling in this case appeared to be substantially reduced relative to that in the element because of significant mixing of $6p_{1/2}$ and $6p_{3/2}$ states when forming the cluster, the largest splitting being ~0.21 eV in the former t_g HOMO. As expected, both the total energy and net overlap populations also decreased relative to the EHT results because of orbital contractions. According to EHT, the distortion from I_h to the observed T_h symmetry, with surface distances of 3.281 to 3.416 Å, generates little energy gain, the change in symmetry evidently arising rather from specific cation effects. The loss of one electron from t_g in Tl_{13}^{10-} has only a minute effect on the average Tl–Tl surface overlap population. Centering and coulomb effects again appear to play the dominant roles in the stabilization of these clusters. A less well isolated cluster has also been found in the semiconducting $Na_2Cd_2K_6(Tl_{12}Cd)$, which occurs in the former Im$\bar{3}$ space group [97]. Here, cadmium not only centers an analogous $Tl_{12}Cd^{12-}$ cluster but also randomly substitutes on ~50% in the unique Na sites, leading to intercluster Cd bridges that appear considerably more covalent. This then is somewhat akin to the intermetallic examples of zinc, etc. noted above.

The unusual prolificacy of these heavy metal clusters, especially for Tl, raises another question about whether the Tr–Tr bonding in them is in some absolute sense unusually strong. This probably pertains only in a relative sense, that is, relative to bond strengths in competitive phases. In the long run, the alternatives to these clusters or networks are the alkali and Ga, In, Tl metals themselves, and these are generally not considered to be particularly strongly bound, especially the former. The enthalpy of atomization of Tl is half of that of La, three times that of Hg.

3.3
Network Structures and New Clusters

The three heavier triel elements additionally form numerous anionic network structures in which many of the building blocks are various recognizable deltahedra, and some of these are larger than presently known in isolated form. Gallium is the most productive of networks and thallium, the least [79, 98]. These phases are germane to the present article in that they both illustrate further applications and the utility of electronic rules for polyhedra and emphasize the general pervasive driving force for closed-shell configurations. The broadest categorization of phases with the last feature is as Zintl phases, and the simplest varieties contain anionic groups or networks of the *p* elements that follow octet rules. The structural and compositional diversity and variability among these is remarkable [5–7]. It has also become appropriate to broaden the Zintl-Klemm concept to include clusters that exhibit nonclassical (delocalized) bonding.

The interconnection of clusters into networks via exo bonds follows from the simple notion that a more-or-less localized *s* pair is located at each vertex of an isolated cluster, energetically lying below the skeletal bonding MO's. Interconnection of clusters by what appear to be normal two-electron–two-center bonds follows a one-electron oxidation of the *s* pair on both clusters and, presumably, some rehybridization or *s*–*p* mixing [98–101]. Thus, members of the isolated *closo*-Tr_n^{-n-2} cluster family become Tr_n^{-n-2+m} when interbonded at *m* vertices to other clusters, or Tr_n^{-2} when they are so bridged at all *n* vertices. This then is yet another means of reducing larger cluster charges.

Only a few examples will be cited rather than a general overview of cluster network phases [98]. Probably the simplest example is the structure of Rb_2In_3 [102] (Cs_2In_3, K_2Ga_3 [80]) shown in Fig. 8. Here classical In_6^{8-} octahedral clusters can be imagined to have been oxidized by $4e^-$ and then joined into square nets by intercluster bonding at the four waist vertices, viz., as $^2_\infty[In_6^{4-}]$. The result is thus a two-dimensional analogue of the CaB_6 structure. The compound is diamagnetic and has a room temperature resistivity above 1000 μohm · cm.

Many other networks structures are complex and seem to involve intricate considerations and delicate balances in order to generate suitable interconnected networks of polyhedra that individually both meet, or come close to, electronic requirements and afford room and good bonding for the necessary cations, all with good space filling. From another viewpoint, the marked complexities of many network structures attest to the importance of valence considerations, even when precise electron balances between apparent cluster network requirements and the cations available apparently are not quite achieved [103, 104]. In fact, the need for precise agreement between these two is probably secondary in general bonding consideration [79].

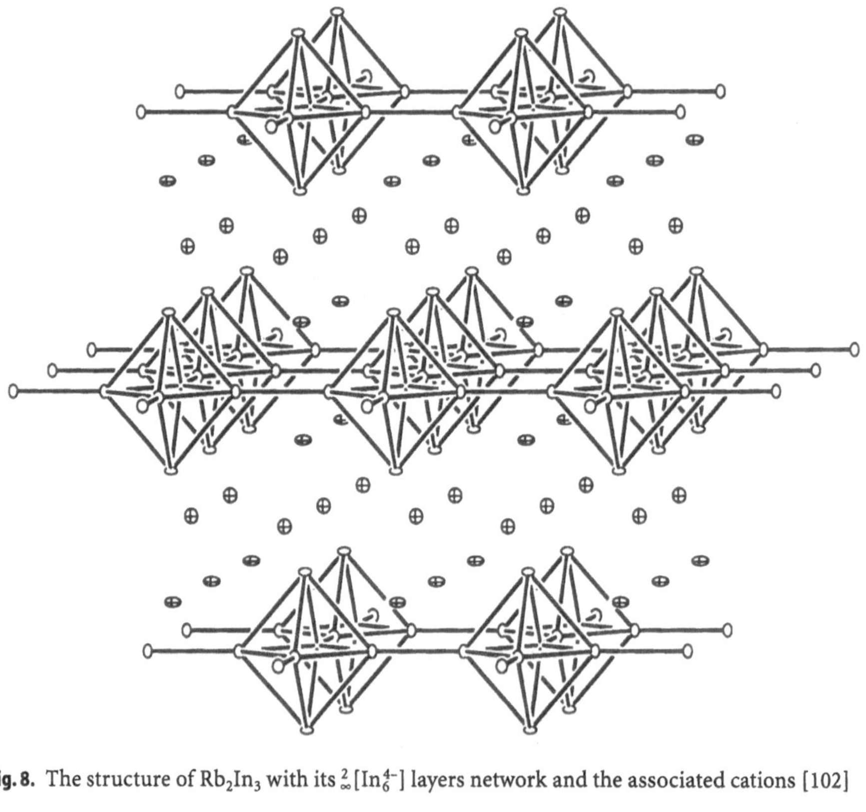

Fig. 8. The structure of Rb_2In_3 with its $^2_\infty[In_6^{4-}]$ layers network and the associated cations [102]

One simple example with new clusters is provided by the $A_3Na_{26}In_{48}$ structure shown in Fig. 9 [94, 105]. Here the A cations, K, Rb or Cs, are located in the larger cavities between the "drums", which are *arachno*-In_{12} with 12 exo bonds. These are interconnected via 12-bonded (12b) *closo*-In_{12} (T_h) icosahedra into a network which is the hierarchical equivalent of the Cr_3Si (A15) type with clusters replacing atoms. EHMO calculations on the former cluster with dummy exo bonds included show the expected $2n+6=30$ (p) electron skeletal count, making the "drum" In_{12}^{6-}. The 12b icosahedra would correspondingly be In_{12}^{2-}. The cation count relative to these polyhedral and network needs show that the phase appears to be 18 electrons long per cell ($Z=2$) compared with the 328 electrons required [2(26+12)+6(30+12)]. However, this difference of 18 corresponds to 31% of the cations per cell, which seems unusual if this large an electron excess exists solely because of a need to fill the annular spaces with 18 extra cations. Perhaps there are electronic requirements of the lattice that are

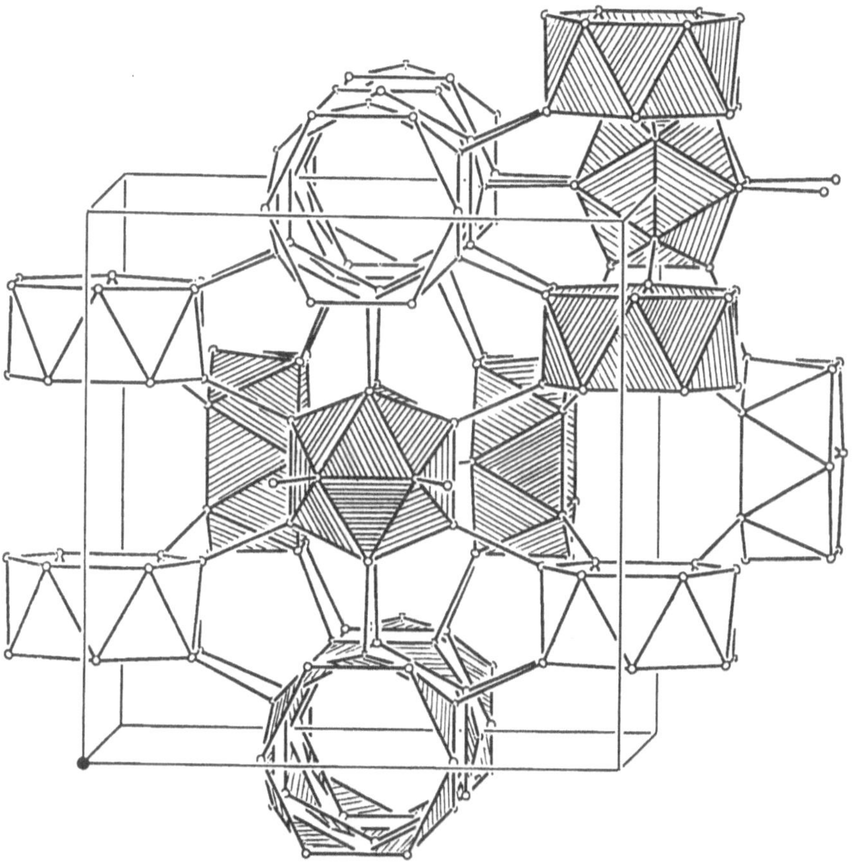

Fig. 9. The network structure of $K_3Na_{26}In_{48}$ constructed from 12-bonded In_{12} icosahedra and 12-bonded In_{12} "drums"

not evident from looking at the pieces; a band calculation has not yet been pursued. (Another description has without comment allocated 12 of these 18 electrons to bonds to the sodium that center the drums [105].)

Gallium forms a number of apparently electron-precise network examples too, although comparative data on properties have not appeared. Several phases incorporate cluster building blocks already described, *closo*-Ga_{12} and *nido*-Ga_{11} as well as a new member, the *closo*-Ga_8^{10-} (D_{2d}). Thus the cell for K_3Ga_{13} (Z = 8) can be described as $(12b\text{-}Ga_{12}^{2-})_4(11b\text{-}Ga_{11}^{2-})_4(3b\text{-}Ga^{o})_4(4b\text{-}Ga^{-})_8$ and for AGa_3 (Z = 3, A = K, Rb, Cs), as $(8b\text{-}Ga_8{}^{2-})(4b\text{-}Ga^{-})$ [98].

Larger polyhedra have also been encountered. One is the *closo*-In_{16} unit shown in Fig. 10. This occurs in $Na_7In_{11.8}$ as an eight-bonded D_{2d} (close to T_d) unit along with In_{11} and In_{10} icosahedral fragments [106]. A similar result is a 12-bonded In_{16} cluster of C_{2v} symmetry in $Na_{15}In_{27.4}$, where variable occupancy on one site in a 10-bonded *nido*-In_{11} is also present [21]. Both are centered by sodium ions. An In-centered In_{16} network component has also been discovered in $K_{17}In_{41}$ [107]. The icosioctahedral In_{16} may be also described as a tetracapped truncated tetrahedron. Calculations on the first In_{16} example above clearly predict that the elongated figure should be closed shell with $2n+4$ skeletal electrons, viz., In_{16}^{20-}. This is one of a group of larger closo clusters that does not obey the $n+1$ bonding orbital rule, with either n or $n+4$ considered possible earlier [108, 109]. Our calculations on these larger polyhedra also indicate that some of the higher lying cluster orbitals occur on the few indium atoms that lack exo bonds; that is, $s-p$ mixing is appreciable at these positions and generates outward pointing lone-pairs. (This has also been noted for networks of larger gallium clusters [92].) In fact, intercluster lone pair repulsions in $Na_7In_{11.8}$ between such pairs on In_{16} and the single vertex in *nido*-In_{11} that is not exo bonded evidently push the latter to an antibonding and therefore empty state. When these separate parts are added up, the refined $Na_7In_{11.8}$ composition and structure appear to contain only three electrons per cell in excess of the 507 required to generate the network observed, probably within experimental error.

Examples of other large polyhedra are also found in networks. However, some contain fractionally occupied positions, and most occur in mixed p-element systems where the locations of the heterometals (Zn, Cd, Au, Sn, etc.) are not always clear. Six-bonded *closo*-In_{16} clusters that are close to T_d symmetry, as above, also appear in a phase $Na_9In_{16.8}Zn_{2.3}$ [110]. But in this case, only about 50% occupancy is found for the four atoms that cap hexagonal faces on the truncated tetrahedron and also have nominal exo bonds. (This is not the case in Fig. 10.) Bonding in such seven-coordinate situations appears to be less favorable, since the exo bond is often replaced by a missing vertex and a lone pair on the other atom. (The combination does not change the required electron count.) Similar atom deficiencies in such capping sites have been seen with $(Ga,Cd)_{16}$ cluster units in $Na_{8.75}Ga_{14.0}Cd_{5.9}$ [111], in a 15-bonded *closo*-In_{15} (D_{3h}) in hexagonal $Na_{23}In_{38.4}Zn_{4.6}$ that can be described as a tricapped truncated trigonal prism, and in a 12-bonded closo-In_{18} D_{3d} unit ($2n+4$) in $Na_{49}In_{80.9}Sn_{9.1}$ that is based on a hexacapped truncated trigonal antiprism [110]. Polyhedra of 15, 18 and 20 atoms in other phases may be better described as "spacers" rather than clusters. Fig. 11 shows a novel unit that appears in two structures in the

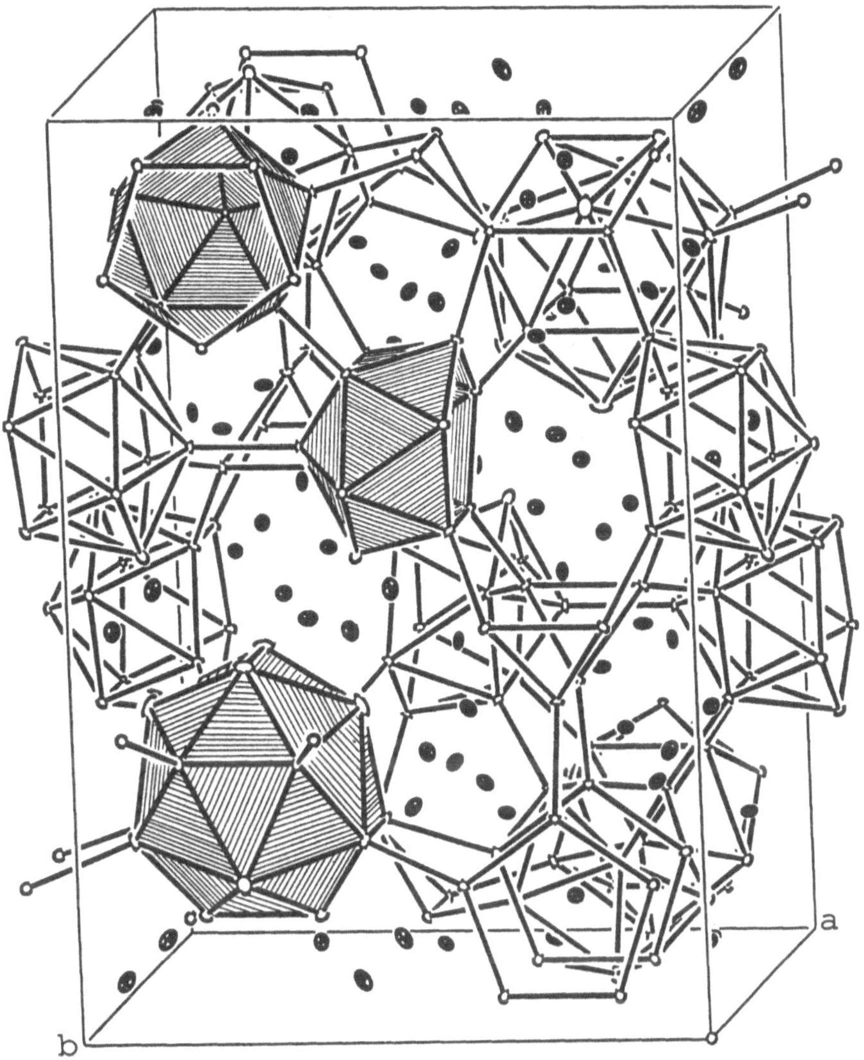

Fig. 10. The eight-bonded *closo*-In_{16} polyhedra in $Na_7In_{11.8}$. Ten-bonded icosahedral fragments and interbonded In_3 units are also evident. All In–In separations <3.5 Å are marked

K–In–Cd system, an *arachno*-$(In,Cd)_{18}$ version of the unknown *closo*-M_{20}, or a "big mac" [110]. A closely related "tubular cluster" $Cd_{12}In_6$ has been found in $Na_8K_{23}Cd_{12}In_{48}$ where it is similarly bonded at the outer cadmium vertices to 12 indium icosahedra [112]. The outer hexagons in these "triple deckers" are often larger in diameter than the central hexagon so that their polyhedral parentage seems less certain.

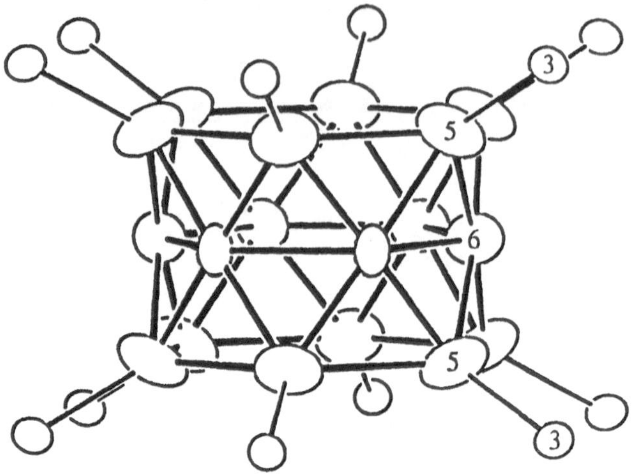

Fig. 11. The *arachno*-$(In,Cd)_{18}$ cluster "big mac" in a K–In–Cd network structure [110]. The 5–5 distances in the outer six-rings are quite long, 3.62 Å

4
Closure

To the writer, the many and diverse manifestations of significant homoatomic bonding among the heavy post-transition (p) elements in so many clusters and related phases is remarkable. That the great fraction of the elements in these examples occur with negative oxidation states is a further wonder when it is remembered how very sparse the same elements are of cationic examples with significant metal–metal (or semimetal) bonding. Notwithstanding, the vast majority of these species fit easily into polyhedral electronic generalities first articulated as "Wade's Rules", or what are reasonable modifications thereof, considerable tributes to both the insight of the approach and the power of these bonding models.

These cluster examples have also taught, or refreshed, some powerful lessons about the coulombic stabilization achievable in "neat" salts. Even cursory looks at the structural chemistry of the alkali-metal systems with Zn and Cd reveals more cluster-like constructions but ones that are more heavily condensed onto one another, presumably because of a greater paucity of electrons, and doubtlessly other things we don't understand or appreciate. Electronic regularities in condensed and fused clusters are generally more difficult to identify [101]. We have also learned features that we were often too unimaginative or near-sighted to anticipate. The chemical ways by which electron-poorer cluster systems appear to be stabilized are good lessons learned, or refreshed. These presently encompass: 1) significant distortion of conventional clusters to ones with higher order vertices, 2) removal of multiple adjacent vertices along with some electron pairs, 3) centering with at least main-group heteroelements, and 4) oxidation and formation of cluster networks. What is now known is doubtlessly still limited by the efforts, imagination, luck and serendipity of the explorers.

Acknowledgements. The author is greatly indebted to an outstanding group of coworkers for their major contributions to some of the discoveries described herein and the excitement these have generated. Our part of this work has been supported by the Office of Basic Energy Sciences, Materials Sciences Division, U.S. Department of Energy. The Ames Laboratory is operated by DOE by Iowa State University under Contract No. W-7405-Eng-82.

5 References

1. Wade K Inorg Nucl Chem Lett (1972) 8:559
2. Wade K Adv Inorg Chem Radiochem (1976) 18:1
3. Corbett JD Chem Rev (1985) 85:383
4. Shoemaker DP, Marsh RE Acta Crystallogr (1953) 6:197
5. Hewaidy IF, Busmann E, Klemm W Z Anorg Allg Chem (1964) 328:283
6. Schäfer H Annu Rev Mat Sci (1985) 15:1
7. Kauzlarich S ed "Chemistry, Structure and Bonding of Zintl Phases and Ions" VCH Publishers: New York (1996)
8. Hershaft A, Corbett JD Inorg Chem (1963) 2:979
9. Corbett JD, Rundle RE Inorg Chem (1964) 3:1408
10. Britton D Inorg Chem (1964) 3:305
11. Corbett JD Inorg Chem (1968) 7:198
12. Corbett JD Prog Inorg Chem (1976) 21:129
13. Bjerrum NJ, Boston CR, Smith GP, Davis HL Inorg Nucl Chem Lett (1965) 1:141
14. Bjerrum NJ, Smith GP Inorg Chem (1967) 6:1968
15. Krebs B, Hucke M, Brendel CJ Angew Chem (1982) 94:453
16. Krebs B, Mummert M, Brendel C J Less-Common Met (1986) 116:159
17. Friedman RM, Corbett JD Inorg Chem (1973) 12:1134
18. Corbett JD, Sevov SC Z Phys D Atoms Molecules and Clusters (1993) 26:64
19. Critchlow SC, Corbett JD Inorg Chem (1982) 21:3286
20. Critchlow, SC, Corbett, JD Inorg Chem (1985) 24:979
21. Sevov SC, Corbett JD J Solid State Chem (1993) 103:114
22. Smith JF, Hansen DA Acta Crystallogr (1967) 22:836
23. Prince DJ, Corbett JD, Garbisch B Inorg Chem (1970) 9:2731
24. Couch TW, Lokken DA, Corbett JD Inorg Chem (1972) 11:357
25. Critchlow SC, Corbett JD Inorg Chem (1984) 23:770
26. Cisar A, Corbett JD Inorg Chem (1977) 16:2482
27. Edwards PA, Corbett JD Inorg Chem (1977) 16:903
28. Dong Z-C, Corbett JD J Am Chem Soc (1994) 116:3429
29. Dong Z-C, Corbett JD Inorg Chem (1996) 35:3107
30. Zhao J-T, Corbett JD Inorg Chem (1995) 34:378
31. Liu Q, Hoffmann R, Corbett JD J Phys Chem (1994) 98:9360
32. Henning R, Leon-Escamilla EA, Zhao J-T, Corbett JD Inorg Chem (1997) 36: 1282
33. Dong Z-C, Corbett JD Angew Chem Int Ed Eng (1996) 35:1006
34. Belin CHE, Corbett JD, Cisar A J Am Chem Soc (1977) 99:7163
35. Corbett JD, Edwards PA J Am Chem Soc (1977) 99:3313
36. Burns RC, Corbett JD Inorg Chem (1985) 24:1489
37. Campbell J, Dixon DA, Mercier HP, Schrobligen GJ Inorg Chem (1995) 34:5798
38. Burns RC, Corbett JD J Am Chem Soc (1982) 104:2804
39. Belin C, Mercier H, Angilella V New J Chem (1991) 15:931
40. Friedman RM, Corbett JD Inorg Chim Acta (1973) 7:525
41. Angilella V, Belin C J Chem Soc Faraday Trans (1991) 87:203
42. Fässler T, Hunziker M Inorg Chem (1994) 33:5380
43. Critchlow SC, Corbett JD J Am Chem Soc (1983) 105:5715

44. Cordier G, Müller V Z Naturforsch (1994) 49b:935
45. Dong Z-C, Corbett JD J Am Chem Soc (1995) 117:6447
46. Dong Z-C, Corbett JD J Am Chem Soc (1995) 34:5709
47. Sørlie M, Smith GP J Inorg Nucl Chem (1981) 43:931
48. Gillespie RJ Chem Soc Rev (1979) 8:315
49. Corbett JD, Adolphson DG, Merryman DJ, Edwards PA, Armatis FJ J Am Chem Soc (1975) 97:6267
50. Kummer D, Diehl L Angew Chem Int Ed Engl (1970) 9:895
51. Burns RC, Corbett JD J Am Chem Soc (1981) 103:2627
52. Eichhorn BW, Haushalter RC, Pennington WT J Am Chem Soc (1988) 110:8704
53. Eichhorn BW, Haushalter RC J Chem Soc Chem Commun (1990) 937
54. Callahan GB, Hawthorne MF Adv Organometallic Chem (1976) 14:145
55. Schiemenz B, Huttner G Angew Chem Int Ed Engl (1993) 32:297
56. Belin CHE private communication (1996)
57. Critchlow SC PhD Dissertation, Iowa State University (1983)
58. Rudolph RW, Wilson WL, Parker F, Taylor RC, Young DC J Am Chem Soc (1978) 100: 4629
59. Guggenberger LJ, Muetterties EL, J Am Chem Soc (1976) 98:X7221
60. Fässler TF, Hunziker M Z Anorg Allg Chem (1996) 622:837
61. Fässler TF, Hunziker M Z Anorg Allg Chem, submitted
62. Lohr LL Inorg Chem (1981) 20:4229
63. Burns RC, Gillespie RJ, Barnes JA, McGlinchey MJ Inorg Chem (1982) 21:799
64. King RB Inorg Chim Acta (1982) 57:79
65. O'Neill ME, Wade K Inorg Chem (1982) 21:461
66. Rudolph RW, Taylor RC, Young DC "Fundamental Research in Homogeneous Catalysis" Tsutsui M Ed Plenum Press: New York (1979), p. 997
67. Wilson WL, Rudolph RW, Lohr LL, Pyykkö P Inorg Chem (1986) 25:1535
68. Martin TP, J Chem Phys (1985) 83:78
69. Wheeler RG, LaiHing K, Wilson WL, Duncan MA J Chem Phys (1988) 88:2831
70. Schild D, Pflaum R, Riefer G, Recknagel E J Phys. Chem (1987) 91:2647
71. Schild D, Pflaum R, Riefer G, Recknagel E Z Phys D–Atoms, Molecules and Clusters (1988) 10:329
72. Hartmann A, Weil K. G Angew Chem Int Ed Engl (1988) 27:1091
73. Sevov SC, Corbett JD Inorg Chem (1991) 30:4875
74. Corbett JD Inorg Nucl Chem Lett (1969) 5:81
75. Zintl E, Dullenkopf W Z Phys Chem Abt B (1932) 16:183
76. Schmidt PC Structure and Bonding (1987) 65:91
77. Belin C Acta Crystallogr (1980) B36:1339
78. Frank-Cordier, U, Cordier G, Schäfer H Z Naturforsch (1982) B37:119, 127
79. Corbett JD, "Chemistry, Structure and Bonding of Zintl Phases and Ions" Kauzlarich S ed VCH Publishers: New York (1996) Chap 3
80. Henning R, Corbett JD, to be published
81. Blase W, Cordier G, Müller V, Häussermann U, Nesper R Z Naturforsch (1993) 48b:754
82. Sevov SC, Corbett JD, Ostenson JE J Alloys Compd (1993) 202:289
83. Cordier G, Müller V Z Kristallogr (1992) 198:281
84. Dong Z-C, Corbett JD J Cluster Sci (1995) 6:187
85. Dong Z-C, Corbett JD Inorg Chem (1995) 34:5042
86. Dong Z-C, Corbett JD Inorg Chem (1993) 115:11299
87. Dong Z-C, Corbett JD Inorg Chem (1996) 35:2301
88. Sevov SC, Corbett JD Inorg Chem (1993) 32:1059
89. Dong Z-C, Henning R, Corbett JD Inorg Chem (1997) 36: accepted
90. Henning R, Corbett JD to be published
91. Sevov SC, Corbett JD J Am Chem Soc (1993) 115:9089
92. Burdett JB, Canadell E Inorg Chem (1991) 30:1991
93. Corbett JD J Alloys Compd (1995) 229:10

94. Sevov SC, Corbett JD Inorg Chem (1993) 32:1612
95. Llusar R, Beltrán A, Andrés J, Silvi B, Savin A J Phys Chem (1995) 99:12483
96. Pyykkö P in "Methods in Computational Chemistry" Wilson S ed., Plenum Press: New York (1988) 2:137
97. Tillard-Charbonnel MM, Belin CHE, Manteghetti AP, Flot DM Inorg. Chem (1996) 35:2583
98. Belin C, Tillard-Charbonnel M Prog. Solid State Chem (1993) 22:59
99. Belin C, Ling RG J Solid State Chem (1983) 48:40
100. Schäfer H J Solid State Chem (1985) 57:97
101. Burdett JK, Canadell E J Am Chem Soc (1990) 112:7207
102. Sevov SC, Corbett JD Z Anorg Allg Chem (1993) 619:128
103. Nesper R Prog Solid State Chem (1990) 20:1
104. Sevov SC, Corbett JD J Solid State Chem (1996) 123:344
105. Carrillo-Cabrera W, Ceroca-Canales N, Peters K, von Schnering HG Z Anorg Allg Chem (1993) 619:1556
106. Sevov SC, Corbett JD Inorg Chem (1992) 31:1895
107. von Schnering HG private communication (1996)
108. Brown LD, Lipscomb WN Inorg Chem (1977) 16:2989
109. Fowler PW Polyhedron (1985) 4:2051
110. Sevov SC Ph.D. Dissertation, Iowa State University, 1993.
111. Charbonnel MT, Belin C Mat. Res. Bull (1992) 27:1277
112. Flot DM, Tillard-Charbonnel MM, Belin CHE J Am Chem Soc (1996) 118:5229

Author Index Volumes 1–87

Banci L, Bertini I, Luchinat C (1990) The 1H NMR Parameters of Magnetically Coupled Dimers - The Fe2S2 Proteins as an Example. *72*:113-136
Baran EJ, see Müller A (1976) *26*:81-139
Bartolotti LJ (1987) Absolute Electronegativities as Determined from Kohn-Sham Theory. *66*:27-40
Bau RG, see Teller R (1981) *44*:1-82
Baughan EC (1973) Structural Radii, Electron-cloud Radii, Ionic Radii and Solvation. *15*:53-71
Bayer E, Schretzmann P (1967) Reversible Oxygenierung von Metallkomplexen. *2*:181-250
Bearden AJ, Dunham WR (1970) Iron Electronic Configuration in Proteins: Studies by Mössbauer Spectroscopy. *8*:1-52
Bencini A, see Banci L (1982) *52*:37-86
Benedict U, see Manes L (1985) *59/60*:75-125
Benelli C, see Banci L (1982) *52*:37-86
Benfield RE, see Thiel RC (1993) *81*:1-40
Bergmann D, Hinze J (1987) Electronegativity and Charge Distribution. *66*:145-190
Berners-Price SJ, Sadler PJ (1988) Phosphines and Metal Phosphine Complexes: Relationship of Chemistry to Anticancer and Other Biological Activity. *70*:27-102
Bertini I, see Banci L (1990) *72*:113-136
Bertini I, Ciurli S, Luchinat C (1995) The Electronic Structure of FeS Centers in Proteins and Models. A Contribution to the Understanding of Their Electron Transfer Properties. *83*:1-54
Bertini I, Luchinat C, Scozzafava A (1982) Carbonic Anhydrase: An Insight into the Zinc Binding Site and into the Active Cavity Through Metal Substitution. *48*:45-91
Bertrand P (1991) Application of Electron Transfer Theories to Biological Systems. *75*:1-48
Bill E, see Trautwein AX (1991) *78*:1-96
Bino A, see Ardon M (1987) *65*:1-28
Blanchard M, see Linarès C (1977) *33*:179-207
Blasse G, see Powell RC (1980) *42*:43-96
Blasse G (1991) Optical Electron Transfer Between Metal Ions and its Consequences. *76*:153-188
Blasse G (1976) The Influence of Charge-Transfer and Rydberg States on the Luminescence Properties of Lanthanides and Actinides. *26*:43-79
Blasse G (1980) The Luminescence of Closed-Shell Transition Metal-Complexes. New Developments. *42*:1-41
Blauer G (1974) Optical Activity of Conjugated Proteins. *18*:69-129
Bleijenberg KC (1980) Luminescence Properties of Uranate Centres in Solids. *42*:97-128
Boca R, Breza M, Pelikán P (1989) Vibronic Interactions in the Stereochemistry of Metal Complexes. *71*:57-97
Boeyens JCA (1985) Molecular Mechanics and the Structure Hypothesis. *63*:65-101
Böhm MC, see Sen KD (1987) *66*:99-123
Bohra R, see Jain VK (1982) *52*:147-196
Bollinger DM, see Orchin M (1975) *23*:167-193
Bominaar EL, see Trautwein AX (1991) *78*:1-96
Bonnelle C (1976) Band and Localized States in Metallic Thorium, Uranium and Plutonium, and in Some Compounds, Studied by X-ray Spectroscopy. *31*:23-48
Bose SN, see Nag K (1985) *63*:153-197
Bowler BE, see Therien MJ (1991) *75*:109-130
Bradshaw AM, Cederbaum LS, Domcke W (1975) Ultraviolet Photoelectron Spectroscopy of Gases Adsorbed on Metal Surfaces. *24*:133-170
Braterman PS (1972) Spectra and Bonding in Metal Carbonyls. Part A: Bonding. *10*:57-86
Braterman PS (1976) Spectra and Bonding in Metal Carbonyls. Part B: Spectra and Their Interpretation. *26*:1-42
Bray RC, Swann JC (1972) Molybdenum-Containing Enzymes. *11*:107-144
Brec R, see Evain M (1992) *79*:277-306